An Evolutionist deconstructs Creationism

An Evolutionist deconstructs Creationism

by Arndt von Hippel

This book was printed in the United States of America.

ISBN: 1-58820-198-8

This book printed on acid free paper.

Library of Congress Control Number: 00-092628

1stBooks - rev. 11/20/00

PREFACE

Creationists contend that similarities between life forms reveal *the thought and touch of our Creator.* Evolutionists insist that life forms resemble each other because *all share the same ancient ancestors.* These contradictory hypotheses cannot both be true. So which does the evidence support?

Well, the creationist claim that *"God created life 'as is - where is' during one hectic week just a few thousand years ago"* rests entirely upon a single anonymous anecdotal report in Genesis. On the other hand, the evolutionist conclusion that *"Life evolved gradually over billions of years"* is supported by all relevant findings in every field of modern science. Thus to sustain their subjective beliefs, creationists must ignore, reject or distort an immense amount of detailed, carefully confirmed, scientific evidence.

The case for life's evolution is familiar to all who paid attention as their high school science teacher went on about the radiometric dating of fossils, continental drift, analyses of ice cores from Arctic and Antarctic glaciers, and even tree ring evidence from wood that grew long before "Creation". The usual creationist response (e.g., in Kansas, 1999) has been to degrade public high school education by limiting student exposure to scientific knowledge that discredits Hebrew Bible (Old Testament) tales. For creationism only flourishes among the scientifically illiterate, the defeated, the logically challenged, and those indoctrinated since early childhood.

In any event, despite 150 years of religion-based assaults upon the sciences that sustain our technological civilization, the explanatory and predictive powers of modern Evolution Theory have become key to innovative advances in agriculture, biology, chemistry, computer science, ecology, geology, infectious diseases, paleontology, psychology and many other fields. So the question naturally arises - might Evolution Theory open the way for a rational investigation of creationist claims to absolute truth, natural law and divine guidance? Surprisingly enough, the answer is "Yes!"

In this **highly readable, thought-provoking, amusing little book,** Dr. Arndt von Hippel develops many original, evolution-based insights into *how, when, where and why religions arise, compete, adapt and die.* He also dissects the doctrinal illogicalities, historical inconsistencies and antiscientific biases of creationism to demonstrate that these theological aberrations are unavoidable side-effects of evolutionarily correct behavior.

The book ends with twenty-two poems and mini-essays that explore matters ranging from funerals, unwed fathers and Pascal's wager on eternity, to the secular nature of morality, the allure of virtual realities and the mysteries of the Holy Trinity.

Table of Contents

Part Two: Religion from A to Z (in verse and brief essays)

Chapter One

****

"Theology is the ignorance of natural causes reduced to a system"

Paul Henri Thiry in Common Sense (1772)

****

Can Evolution Theory provide Theological Insights?

Modern Evolution Theory is the only scientifically tenable explanation for life on Earth or anywhere else. It was initiated in the late 1850's by Darwin and Wallace, who independently recognized that life's incredible diversity and adaptibility were unavoidable consequences of excessive reproduction and simple variation. For when prolific ancestors produce many prolific descendents, the resulting fierce competition over limited resources will be won by those best suited to local conditions. Thus the next generation automatically inherits the adaptive traits of prolific survivors.

Similarly, because relatively few domesticated plants or animals are selected to reproduce, their desirable qualities soon spread throughout the entire cohort. Of course, if only the best-adapted or "best quality" individuals survive to breed, the ensuing gain in average population fitness simply raises the bar that subsequent generations must clear. And as changing times open additional opportunities for previously irrelevant variations, those opportunities and incremental adaptive changes coalesce into new species without any need for godly input or interference.

Darwin and Wallace were led to their evolutionary hypothesis by a great deal of evidence. They realized that mountains rose slowly as a result of earthquakes and volcanic activity. They understood how gradual erosion by water or wind could bring about massive changes over millions of years. And they were impressed by the extensive and varied fossil record, with its grand displays of primitive and long-extinct life forms.

Furthermore, each of these widely traveled naturalists had encountered an abundance of living species previously unknown to European or North American scientists. Both had also described "new" species that existed only on specific, recently formed, volcanic islands. Yet even those recently placed plants and animals appeared "especially designed" for their novel surroundings - an outcome which could only have come about through sweeping multigenerational adaptive makeovers of distantly related mainland forms that had been cast upon these isolated shores long before.

Of course, both men were acutely aware that all **their evidence for Earth's great antiquity - and for ongoing new species formation - contradicted the widely accepted creation story of the Hebrew Bible.** Hence they only presented their case cautiously, after extensive study.

Nonetheless, their proposals were necessarily incomplete. For science still had many advances to make before the genetic basis of inheritance - or even the existence of molecules - would be established. So while their guesses about the specific mechanisms of biological evolution have largely been superseded, the original observations and general conclusions of Darwin and Wallace were remarkably prescient. And they remain valid to this day.

Evolution is still a disturbing idea

Although the disturbing idea of biological evolution was presented with circumspection, it spread rapidly upon release. Not surprisingly, it shocked many who had been taught that God created a fixed number of animal species - along with man in His Image - during a particularly busy week some thousands of years earlier. In addition, because evolution theory redefined life as a non-miraculous process - and implied that man was merely another animal - it drastically diminished God's responsibilities and relevance.

But in the end, the enormous cognitive leap from Bible-based Creationism to evidence-based Evolution Theory was powered by simple observations that anyone could make in their own back yard, such as:

Nature is prolific. Resources are limited. Competition is fierce. Variation is inevitable. Only the most fit survive to reproduce. Thus natural selection brings about endless adaptive changes.

It has been 150 years since Darwin and Wallace drew the world's attention to the infinite originality and creativity of selective incremental and iterative changes. In that time, evolution theory has contributed to innumerable advances in almost every field of science - from computer design and molecular biology to psychology, geology and agriculture.

So could evolution theory possibly cast new light upon such an ancient and subjective cultural tradition as religion?

Well, actually, yes. Consider the following ordinary observations:

Older religions give rise to descendents that somewhat resemble them.

Successful religions differ from their ancestral forms in important ways.

New religions lack historical acceptance. Their advantage lies in the rapid creation of fresh dogma and easy deletion of outmoded ancestral doctrines.

Religious resources (believers and their assets) are always limited.

Therefore, all religions must compete fiercely to survive and prosper.

Prosperous religions produce more descendent religions.

Times change, and religions adapt or die.

In other words, regardless of natural law, absolute truth, or their ultimate provenance, religions arise, adapt and die in strict accordance with modern evolution theory, for:

Many religions are born but only a few mature or leave descendents.

New religions seek comfort and legitimacy by resembling their ancestors.

New religions achieve prosperity by offering something different.

Wealthy older religions enjoy conservative leaders and outdated ideas.

Hence older religions gradually become unresponsive and irrelevant.

Outmoded religions are eventually displaced by descendent forms more in tune with (better adapted to) the current social reality.

Religions make sense from an evolutionary perspective

The above observations suggest that ***current social reality*** limits the design and marketing of any new religion, and that ***ultimate reality*** is merely a figure of speech. Consequently, Evolution Theory appears more relevant to the study of religion than the skimpy theological background of an average evolutionist might suggest. For at least in principle, the same evidence-based deductive approach that solves a crime - or determines whether dinosaur feathers served as insulation before becoming adapted to flight - can also critique the puny flight muscles of angels or cast doubt upon their ability to hover without raising great clouds of dust.*

* For another example of "Virtual Reality", see page 145.

Indeed, when we apply evolutionary logic to certain fundamental, yet seemingly simplistic or even counter-intuitive, religious stories, it becomes clear that those tales actually represent carefully crafted solutions to specific historical problems. And while religions basically market magic and mirage - which ought to exempt them from all common sense constraints - religious doctrines nevertheless undergo repeated revisions in an unending effort to provide the internally consistent rules, regulations and promises that can make a shared delusional system seem more plausible and attractive.

Our modern evolutionary approach would have interested Blaise Pascal (1623-1662), a devout French philosopher and mathematician, whose profound frustration with Christian dogma led him to write "Christians profess a religion for which they cannot give a reason".*

* See also Pascal's "Why not believe?" formula on page 114.

Chapter Two

****

"Descended from the apes! My dear, let us hope that it is not true, but if it is, let us pray that it will not become generally known."

(remark attributed to the wife of the Anglican Bishop of Worcester)

****

Religions are the living fossils of Stone Age science

Humans live in groups because solitary humans cannot prosper, defend themselves or provide adequate child care. However, from the moment that it forms, every group faces internal and external challenges. And sooner or later, environmental changes or unavoidable conflicts over resources bring about yet another violent disruption of traditions, beliefs and life styles. Indeed, human history reveals little more than an appalling sequence of bloody military campaigns and drastic cultural revisions, with stronger or better equipped peoples devastating the weaker.

Overall, the stability and competitive success of any human group is largely determined by its resources, environment, traditions and beliefs. Shared traditions and beliefs preserve useful knowledge, expedite communication, encourage cooperation and sustain behavioral norms. Some simple folk tales become traditional truths through endless retelling. Other beliefs arrive full-grown on the wings of force and chaos. But whatever their origins, all social traditions and beliefs remain dynamic constructs that grow, fade, change and recombine within the greatly embellished story that identifies each people to themselves.

On the usefulness of hypothetical entities and events

Our more successful Stone Age ancestors made great efforts to understand significant phenomena or unusual occurrences so they might develop a reasoned response. Memorable narrative-type explanations tended to persist and evolve as long as they seemed useful, regardless of their plausibility. Of course, ordinary evidence-based explanations required regular revision and updating as well. Thus humans developed, modified and discarded innumerable hypotheses about life and the world.

Many theological constructions failed to attract interest. Some had inadequate support, others offered little that was new or useful. Only a few religious hypotheses grew up to become major religions that replicated successfully in many minds and lasted for centuries or millenia. Yet even the most successful religions were destined to die - often through decisive military defeat by forces claiming the support of more powerful deities.

Polytheism allowed man to negotiate with nature

Intelligence is the ability to evaluate and respond appropriately to various objects and aspects of the environment. As human language developed into the world's best information processing tool, it came to dominate the investigation, categorization and manipulation of complex relationships. Because primitive human societies were totally immersed in wilderness, the most routine human interactions with other creatures - or with notable natural features such as flooding rivers, fearsome forests or storm-lashed mountains - or with erratic phenomena such as earthquakes or fiery volcanoes - could profoundly and immediately affect survival.

Their familiarity with death left primitive peoples pondering personal losses. Important events recurred in dreams - dead people or animals returned to disturb restless sleepers - the world teemed with hypothetical agents bearing unknown potentials for good or evil. At the same time, members of simple communities had little cause to anticipate that the spirit or soul of a departed hunter or peasant would have a joyful afterlife. In fact, it was widely acknowledged that if spirits of the dead did persist, they would remain sequestered in the underworld and resemble those miserable grey shades of formerly robust selves that so often reappeared in dreams.

But as settled groups enlarged and economies developed, it became profitable to predict that the spirits of dead royalty or wealthy citizens would receive preferential treatment - and even retain the use of important objects, animals and personal attendants that were buried with them - as long as generously subsidized priestly negotiations with powerful other-world entities paved the way.

Every culture developed its own names, ranks and categories of deities and spirits. Some deities had special talents for warfare, others reveled in romance or the regulation of comets, volcanoes, wild rivers or stormy seas. The eventual religious consensus reflected the public's appraisal of widely discussed, highly ambiguous predictions by different priests and oracles as these demonstrated their psychological or medical or meteorological or volcanological prowess and marketing skills.

Reasonable persons soon recognized the importance of bribing the human representatives of important deities before insoluble problems arose. However, those deities - and the powerful phenomena that they controlled - frequently remained uncooperative or unpredictable despite generous donations and sacrifices. Yet when gods or goddesses were blamed for evil happenings, this merely brought their human handlers greater credibility and more donations. For the wise hoped to stay on good terms with as many powerful anthropomorphic deities as they could afford, since second or third opinions were occasionally helpful and a person might neutralize an adverse divine intervention by turning one deity against another.

Although the most widely revered polytheistic deities had different names in different cultures, their unique subspecialties made them easily recognizable by outsiders. Internationally revered deities proved especially useful, since they allowed traders and warriors to make binding agreements and swear oaths calling down divine retribution upon anyone who cheated.

Even unfamiliar local deities received respect and tribute from travelers hoping to ensure a safe passage. And while none dared ask for refunds, polytheistic priests had plenty of plausible explanations available for times when things failed to work out.

"You misunderstood our (ambiguous) advice" or "You gave less than your enemy so what did you expect" or "She (the goddess) fell in love with your better-looking opponent". Every failure could be seen as another lesson on the need to give more, and more often. Thus polytheism made sense. It explained the inexplicable. It allowed men to bargain on important matters with powerful and unpredictable, yet understandably anthropomorphic deities. Furthermore, polytheism was profitable. And public expectations were reasonable. In fact, all appeared to be well. *But when polytheism was rudely challenged by monotheism, those gods and goddesses failed to respond.*

Only the fittest religions survive

Rational people do not create and maintain complex physical or mental structures unless anticipated returns greatly outweigh the effort and risk involved. And clearly it requires a powerful long-term commitment to develop and promote a successful new religion - along with a thick skin or strong arm to absorb or return the abuse that such a project inevitably attracts from the competition. Nonetheless, the durability of established religions, and the high birth-rate of new religions, regularly reaffirms that even garage-based religious start-ups may richly reward early investors.*

*See also "Redemption" on page 129.

However, before we can determine who gains what from whom within which religion, we must first decide whether god(s) are real - as in actual, regularly visible and visibly awesome. After all, if god(s) were routinely encountered during ordinary business hours, they would certainly have to be taken into account - "Will this payment be by cash, check, credit card or miracle?" - that sort of thing. Indeed, all-too-powerful god(s) could pretty much take and do as they pleased. And the cringing fearful humans that so slavishly served them would simply have to hope that these awesome entities might not prove too cruel or perverse.

Under such stressful circumstances, some sort of religious finishing school might finally have practical value, especially if directed by a minor god who could teach humans how to behave more divinely - or at least well enough to minimize the risk of lightning and similar signs of heavenly displeasure. Quite possibly, any humans who could not avoid regular encounters with deities would also benefit from lessons on how to kneel, bow, pray, kiss the ring, wash the feet, submit gracefully, and never, never, ever mention compensation or overtime.

On the other hand, intelligent humans should surely resent the forced lack of autonomy, despise their own subservience, and detest those divine rulers - just as reasonable deities ought to find insistent supplicants and practiced cringers incredibly annoying. In short, life on Earth, or any conceivable afterlife, would be living hell for humans who had to spend any time - let alone endure eternity - with actual gods or goddesses. After all, if fleeting fame is enough to make ordinary mortals intolerable, why would anyone wish to interact with perpetually spoiled, far-too-powerful deities?

Regarding the plausibility of religious evidence

So we should thank our lucky stars that deities are so rarely encountered, and be grateful that there has not been a confirmed sighting of any god or goddess for thousands of years. Indeed, if it eventually turns out that supernatural beings once existed, a reasonable person still might conclude that they are now as extinct as the Dodo. But while a dusty Dodo skin, adequate for DNA sampling, may yet reside in some musty museum drawer, you will never encounter a single Divine Pubic Hair, Glorious Fingernail Paring, Angelic Feather Tip or Holy Turd in any cathedral, museum or marketplace.

Thus the deeper we delve into this murky matter of god(s), the more it appears that they have shared the lamentable fate of the Roc - a fabulous bird of enormous size and strength that was once the subject of great speculation. Recent efforts to explain the Roc's persistent absence have focused upon the likelihood that one of its wings was slightly shorter than the other. For during a particularly blinding sandstorm, that unfortunate asymmetry could have caused the Roc to fly in ever-diminishing circles until - with a horrible shriek - he disappeared up his own tailpipe.

This explanation sounds convincing, though it could well be untrue - despite several camel drivers who swore that they heard exactly the right sort of horrible shriek during an especially severe sandstorm. But it certainly implies that there may be no way to prove something totally impossible. Nor, until strict legal limits can be placed upon the human imagination, will anyone ever prove their own innocence of a crime.

Therefore, the best possible verdict for a criminal defendant is never "Innocent!" but merely "Not Guilty" - as in "Guilt Not Proven". So it remains conceivable that a single god or even multiple gods might exist. And as long as tooth fairies continue to serve their purpose, they too will remain conceivable - yet equally unlikely - as in "Existence Not Proven".

What really happens when gods fall out of favor?

Imagine a vast pantheon capable of displaying all the millions of gods and goddesses who have ever been worshipped. Next consider the fact that practically every one of those deities subsequently fell out of favor and was banished from everyday human memory and concern. It might then be logical to ask, "Well now, did any of these gods ever exist?" Indeed,the prior or current reality of any god who has fallen from favor does seem a bit improbable, if only because a real deity could so easily demolish someone who openly defied Him or foolishly declared Her non-existent. "NON-EXISTENT, EH? I'LL SHOW YOU WHO IS NON-EXISTENT!!!"

Apparently even deities are not forever. They too must pass through a cycle of invention, marketing, acceptance and devotion - followed by the inevitable collapse of their credibility when they fail to meet some direct or overwhelming challenge. Of course, these former gods might still exist but simply have no further interest in us and whether we worship them or not - but in that case, their unseen and non-interactive "presence" would be as irrelevant to us as our opinion of their existence has become to them.

Nevertheless, the fact that nobody bothers to worship them anymore strongly suggests that every single god and goddess on the scrap heap of previously revered deities is, and always was, false - just another figment of someone's fertile imagination. Which, by extension, casts strong statistical doubt upon the reality of any currently advocated or future god(s) as well. It therefore seems apparent that god(s) would not remain a regular topic of conversation if certain people didn't keep bringing them up. So who keeps bringing god(s) up, and what is in it for them?

Well, it is clear that a lot of folks make a decent tax-free living (and many exceed their wildest dreams) as a direct result of speaking frequently about, or claiming to speak for, one or more gods. Nonetheless, advertising alone cannot explain the public's desire to consume Coca Cola, especially after all novelty and tooth enamel have worn off. In addition, recovering Coca Cola addicts and recovering Catholics truly seem to miss their former habit or addiction.

So one might ask, "Why are we so easily habituated to cokes and god(s)?" Or more usefully, "What does religion offer to the consumer and what do religious consumers really miss upon losing their faith?"

How about *"A warm social organization - togetherness - a sharing of emotions - a feeling of belonging - a sense of time, place and purpose"?*

Religions can enhance reproductive success

But if the above-mentioned touchy/feely/togetherness type things actually make people feel good, why are people so often ornery and ready to fight? Well, aggression clearly brings reproductive advantage to those who display it in appropriate amounts. That is why aggression routinely reappears in subsequent generations - and why even members of the same family frequently find it hard to get along.

An old Arab saying goes, "It is me against my brother, my brother and me against our cousins, our cousins and us against..." and so on. The point here is, people are damn hard to get along with, they are even harder to convince, and they are almost impossible to organize. Furthermore, the larger any group grows, the more difficult it becomes to get and keep that group coordinated.

But all groups require some structure, for a disorganized group of people is easily overwhelmed by a similar-size but better organized group. ***And this is where religion and patriotism - both justly renowned as the last resorts of scoundrels - gain increasing importance as principles around which societies can rally and cooperate.*** So now we have both a motive for promoting religion - namely, "that so many do so well by doing good" in this fashion, and a "Darwinian" survival benefit for those groups that allow religion to dominate their lives.

Chapter Three

****

"Reason and Ignorance, the opposites of each other, influence the great bulk of mankind. If either of these can be rendered sufficiently extensive in a country, the machinery of government goes easily on. Reason obeys itself; and Ignorance submits to whatever is dictated to it".

Thomas Paine in <u>The Rights of Man</u> (1791)

****

Polytheism begat Monotheism

With a little experience - and in the right hands - monotheism developed into the most powerful way to organize a group that the world had ever known. From the start, monotheism encountered surprisingly little coordinated resistance as it challenged, undermined, discredited and displaced the well-established but decentralized polytheistic free-enterprise system - a system based upon multiple franchises and countless self-authorized affiliates of local, regional or multinational deities - with each successful representative issuing ambiguous advice and accepting tax-free tribute on behalf of a particular deity.

Monotheism's operational strength arose from its unprecedented centralization of authority and resources. For monotheistic priests and secular authorities rapidly recognized the advantages to be gained through cooperation. After all, what king didn't want the additional security and legitimacy of having his rule affirmed by divine appointment? And what religious organization wouldn't jump at the chance to become a sole source provider of heavenly perks? So following an initial bit of floundering, monotheism soon evolved into a powerful divine monopoly that permitted priests to legally extort God's tribute from every soul in the land while divinely appointed kings abused God's preapproval of their credibility by waging war and levying taxes without restraint.

Monotheism organized the group more effectively

Polytheism implied an unavoidable dilution of divine and secular authority, with both continuously up for grabs, as in "May the most powerful god, or the best man, win". In addition, there was always the risk that some major god or goddess might undertake an adverse action or issue a conflicting second opinion that could undermine the high and mighty. And because members of a group often had little reason to rally around a particular king, that official lived in constant dread of losing his head.

On the other hand, by supporting and promoting monotheism, a ruler was likely to gain and retain unprecedented power - especially if he or his cooperating priests succeeded in positioning themselves as sole designated intermediary between Almighty God and members of their group. For such a heavenly appointment automatically authorized the ruler and/or priests to oversee and control every aspect of human behavior - public or private.

Indeed, by eliminating every other source of divine second opinions, ***the opportunity to rule on behalf of a single divine authority made fervent monotheism the final common pathway to ongoing military success and totalitarian rule.***

Absolute power corrupts absolutely

Egyptian monotheism and Hebrew monotheism both represented revolutionary offshoots of Egyptian polytheism. Hebrew monotheism originated after the rise and fall of Egyptian monotheism so the Hebrews derived many useful lessons from that brief Egyptian experience. In turn, the

Christian and Islamic offshoots of Hebrew monotheism put their own unique spins upon traditional Hebrew practices and beliefs. Because Islam was the youngest of the great monotheistic faiths, Mohammed managed to avoid many of the pitfalls that so plagued his Christian predecessors.

Yet no matter how monotheistic dogma and practice might evolve, its essence remained ***a single divinely ordained central authority to which every citizen and believer was directly responsible.*** However, since each of these competing systems interpreted God's Word quite differently, they often found it necessary to defend, promote or enforce their particular hypothesis militarily. Indeed, monotheistic rulers generally behaved as aggressively as current resources would permit.

A significant enhancement of monotheism's military might arose through its ability to convince true believers that those holding different opinions represented an intolerable threat to the One True Faith. Or if this seemed implausible, that God wanted His faithful to save the innocent souls of subhuman pagans - regardless of whether those pagans wanted to be saved. Or if this appeared too preposterous, that pagans might as well be converted and/or killed whenever possible in order to make their assets available for priestly/kingly use, or to do away with dangerously divergent ideas.*

*See also "Belief is a Costly Crutch" on page 109.

Thus the externally destructive, internally corrupting influences of ***absolute power*** and ***conflict of interest*** implicitly or explicitly provided the guidelines for monotheistic operations and doctrines. In addition, the various Hebrew, Christian and Muslim factions routinely heightened religious faith and fervor by making fiery but mutually incompatible claims about the divinely ordained superiority of their particular membership and beliefs.

Furthermore, all monotheists learned from earliest childhood to despise outsiders - including true believers of other monotheistic faiths - and to view all pagans as subhuman, regardless of any fine qualities they might appear to possess.

Monotheists of every persuasion were also taught that their particular faction or sect had a monopoly on ethics and morality. Therefore, while they ought never forget nor forgive wrongs done to their group by others, they were blameless for the inexcusable crimes committed by their own group, since these simply reflected human frailty or God's mysterious will rather than any systemic defect in religious interpretation, methodology or doctrine.*

*See also "Ethics and Moral Behavior" on page 124.

Until recently, it was standard practice for monotheists to threaten and try to kill any members who publicly criticized the local version of God's own truth. This tendency still endures in nations run by monotheistic governments such as Iran (e.g., in the case of Salman Rushdie), Afghanistan, Pakistan, Saudi Arabia and so on. For if someone well-versed in the faith can reject any part of the whole system with impunity, others will soon take similar liberties - until not enough divine truth remains to justify a totalitarian state.

Even where such criticisms occur infrequently or appear inadvertent, any failure to punish severely might make the local version of God's own truth seem less compelling. So while *a just and loving God* is always a great help in marketing, *a fearsome and vengeful God* is what finally sustains any monotheistic fundamentalist ruler.

Similar trickle-down logic leads to "Spare the rod and spoil the child!" Or as the large adult who hits a small child for misbehavior explains, "I am not angry. I do this for your own good because I am a just and loving father. And besides, it hurts me more than it does you!" - which is sometimes true if we factor in parental guilt.

Monotheism: The Original Plan

Near the middle of the second millenium B.C.E., Pharaoh Akhenaten of Egypt summarily ordained that all Egyptians would henceforth and forever worship only the One True Sun God - and also His Only Son (who by fortunate coincidence was none other than Pharaoh Akhenaten himself). In essence, Akhenaten challenged the entire polytheistic medley of Middle Eastern Gods and Goddesses to do their worst. He also defaced their statues and knocked them off their pedestals. And when those disrespected deities failed to strike back, Pharaoh Akhenaten became leader of the world's first monotheistic religion.

This tumultuous birth of monotheism represented far more than just another disagreement over "One God versus Many Gods and Goddesses" in some remote backwater town. For the Egyptian Pharaoh was a big player in old-time Middle Eastern politics. And Akhenaten's insightful self-conversion to self-worshipping monotheism greatly enhanced his temporal power by making him sole beneficiary of all donations, sacrifices, duties and taxes dedicated to God as well as those owed to Pharaoh.

Consequently, this was a truly important theological dispute. And like all truly important theological disputes, the main quarrel was over wealth and power and who would run an extremely lucrative theological enterprise.

So when Pharaoh's sudden conversion to monotheism redirected Egypt's entire discretionary income stream toward himself, the powerful and equally greedy priests, oracles and other religious hangers-on were sorely discommoded.

Akhenaten's insistence that all Egyptians immediately abandon their long-established polytheistic beliefs upset countless conservative polytheists as well. So how did Pharaoh manage to carry it off? And why did he choose to worship the Sun?

Well, undoubtedly, he prepared carefully. And surely surprise was important. But recent studies of growth alterations in ancient tree rings - as well as extensive analyses of ice cores from the Greenland and Antarctic ice caps - all suggest that Earth suffered some great natural calamity at about this time.

That disaster may have been the massive volcanic explosion of nearby Santorini. Or possibly it was a heavy fall of cometary debris during some frightfully close cometary fly-by that folk tales from many lands still describe as a flaming sword (the comet's tail) or fire-breathing dragon stretching across the heavens.*

*See Exodus to Arthur by *Mike Baillie.*

In other words, Akhenaten may have become Pharaoh of a recently devastated land - perhaps at a time when residual atmospheric ash or dust had reduced solar warmth sufficiently to cause widespread weather alterations and famine. Under such circumstances, the rapid return of full-strength sunshine would surely have been a major concern for all survivors.

And since great natural catastrophes in those days were widely interpreted as a withdrawal of Heaven's mandate from the kings or emperors of affected lands, the former pharaoh and/or his priests may have lost their livelihoods and even their lives - for when truly drastic measures appeared necessary, even a ruler might be sacrificed to appease the gods.

Or maybe Akhenaten's sudden conversion to Solar worship sufficiently diverted the people's attention to prevent such an outcome for himself. And because natural disasters sometimes recur, Akhenaten made great efforts to show that his new regime and religion really were different - going so far as to construct a lovely new capital city at Amarna, dedicated solely to worship of the Sun.

But no matter whether Egyptian monotheism arose in response to environmental calamity or divine insight, it lasted only two decades before Pharaoh Akhenaten's natural death - which instilled little additional confidence amongst his loyal followers - brought the experiment to a close.

Shortly thereafter, with the usual back-stabbing process of pharaonic succession still under way, polytheistic priests regained control. At that point, Amarna was not only abandoned but purposely obliterated. And because Pharaoh Akhenaten's Sun God never smote those who desecrated His Holy City of Amarna, the priests had little difficulty bringing ordinary

Egyptians back to polytheism. But Pharaoh's challenge - and the failure of polytheistic deities to respond - did not pass unnoticed.

How do we know? The Bible tells us so

The Hebrew Bible (also known as the Old Testament) is a prime example of a defining social myth. From a superficial point of view, it simply describes the origins, trials and tribulations of the Israelites. At a deeper level, it documents a tortuous progression from widespread polytheism to exclusive monotheism - from the time when a number of religious deities were widely revered by many different peoples, to a time when God had His Chosen People differentiate themselves from the regionally dominant Egyptian form of polytheism, declare it idolatry, disparage polytheists as pagans, sacrifice to Him the young bulls that Egyptians worshipped, and abominate ordinary Egyptian dietary practices such as meat boiled in milk.

Does this sound unnecessarily aggressive? Or was it just critically important that the Israelite priests launch Hebrew monotheism with an attention-getting opening salvo, so all interested parties would simultaneously learn of monotheism's outrageous challenge to every polytheistic god and goddess? After all, by challenging those powerful and well-integrated deities, the unruly Israelites were surely asking for trouble. And their message was clear. "Our God is the only True God! Tell your gods to do their worst. Or admit that they are false!"

Polytheists were naturally concerned when their fearsome deities did not swiftly destroy those presumptuous monotheists. Nevertheless, a great many conservative Hebrew polytheists saw no reason to disown so many beloved gods and goddesses - or abandon their inspiring ceremonies and sensuous festivals - merely to worship an unreasonable God who could not even be depicted. For what intelligent person would voluntarily give up the tried and true for something so new and untested?

On the other hand, there was no need to fuss over it. Let monotheists worship as they pleased, and sooner or later it would all become clear. In the meanwhile, polytheists could continue to respect each other's deities and beliefs as they always had done - even while fighting to steal each other's assets.

However, monotheism's initial credibility - and its only excuse for centralizing assets and authority - arose from the obvious inability of powerful polytheistic gods and goddesses to destroy or even harm the unruly and disrespectful Hebrew followers of the One True God. And with Almighty God having just bluffed His way to an apparent first round victory, this was no time for that hypothetical Hebrew deity to tolerate challenges. So the Israelite priests came up with the most convincing religious argument anyone had ever heard:

"Accept our ideas about God or we will kill you!"

In a word, they invented religious intolerance. And at his first opportunity, Moses had huge numbers of persistent Hebrew polytheists put to death in God's name. Undoubtedly, this terrible massacre encouraged many undecided Hebrews to at least "lip synch" a sincere belief in One God - regardless of whether they truly believed in Him or merely were afraid of the daft ones that did. So with all Hebrew polytheists dead or discreetly silent, the newly unanimous affirmations of faith in One Almighty God soon made that novel proposition seem increasingly familiar and plausible.

The Bible tells us that Moses knew monotheism would not be welcomed by the Hebrews or accepted by the Egyptians. As a result, early monotheists scrambled to provide convincing authentication stories and attractive replacement festivals based upon newly devised or radically revised historical myths. Small wonder that so many of those tales now seem awkward and illogical. Nevertheless, as the only official record of God's Acts, Communications and Commandments, these Hebrew Bible stories defined and legitimated the Israelites while their religion evolved into the world's first successful monotheistic belief system.

The complex historical myths and events that eventually underlay and supported monotheism could only be created slowly and in stages - with repeated market-testing and multiple revisions over centuries until the final effective form was achieved. For example, Hebrew priests may initially have floated God's Second Commandment - which prohibited the worship of anything carved or depicted - to avoid a return of peripheral idol-based sites that might divert worshippers and donations away from the Temple. And as their ideas evolved, the Israelite priests eliminated other weaknesses of Egyptian monotheism as well.

In the end, Hebrew monotheism turned out to be far more resilient and durable than Akhenaten's revolutionary belief system because, among other things, it never rushed historical developments as Pharaoh had been forced to do when establishing monotheism to worship himself in his own lifetime. Anyone familiar with Pharaoh's experiment could easily see why the God of the Israelites would decree forty years of wandering - and a whole new generation to be born, brought up and closely monitored out there in the desert - before the Hebrew nation was founded.

*Of course, with Evolution Theory and the concept of gene-based inheritance yet to come, **the Israelite priests were under no pressure to explain how their derivative religion could possibly be true when its ancestors were all false.***

Chapter Four

****

"...though illusion often cheers and comforts, it ultimately and invariably weakens and constricts the spirit...

"But ...don't exhaust yourself by jousting with religious magic: you're no match for it. The thirst for religion is too strong, its roots too deep, its cultural reinforcement too powerful."

Irvin D.Yalom in Love's Executioner

****

The Hebrew Bible has three critically important parts: Genesis and the Flood, Abraham and Moses

Perhaps Israelite leaders saw monotheism as their only chance to organize the fractious Hebrews into an effective fighting force. Maybe certain priests hoped to institute a more profitable version of monotheism for themselves. Possibly a secular leader convinced some polytheistic Hebrew priests to go along by promising them and their descendents top billing in a carefully designed hereditary priesthood. In any case, the obvious first step - apparently foolhardy, and therefore a great attention getter - was to discredit all those powerful polytheistic gods and goddesses by issuing a rude challenge that none would meet.

Of course, Pharaoh Akhenaten had already shown this first step to be safer than it seemed. But for their entire project to succeed, the Hebrew priests also needed to:

1) *Fabricate an epic history for monotheism that extended back to the very beginning of the universe (confirming Almighty God as the only one).*

2) *Provide God with a mysterious new (and this time, non-solar!) identity.*
3) *Explain some well-known local puzzle, such as those huge fossil bones buried in the sand (Noah's flood eliminated giants from early Earth).*
4) *Decide how old Earth must be (under three thousand years would do).*
5) *Greatly extend life spans on the early "who begat whom" list to keep that list manageably short (as usual, gross exaggerations identify filler material).*
6) *Have God relate Earth's prehistory to someone face-to-face (see Moses).*
7) *Create a convincing special relationship with God so the Hebrews would still worship Him and support their hereditary priests during bad times.*
8) *Have God give the Israelites clear title to good land (see Abraham).*
9) *Let "We were just following God's orders" excuse the nasty manner with which they took possession of that land (see Joshua).*
10) *Have God lay out all details of essential rituals and ceremonies so priestly duties could never be challenged or abolished.*
11) *Maintain the flow of donations (no idols, be kind to poor travelers, make excuses for God during bad times, kill all disbelieving Hebrews, etc).*

Genesis gave our Bronze Age ancestors a popular and entertaining creation myth that is demonstrably untrue in every scientific and historical particular (see below). Nor is there one jot of scientific evidence for a world-wide Flood, despite recently discovered signs suggesting that a sudden breakthrough of salty Mediterranean water catastrophically inundated the Black Sea area less than 7000 years ago - which surely would have provided some surviving farmer named *Noah* with a fearful flood story to pass along.*

*See also "Just Another Stone Age Fable" on page 133.

The story of *Abraham* now seems an embarrassingly poor excuse for that bloody Israelite invasion of Canaan. After all, the oft-repeated claim that "God promised this land to us forever, once we kill the current residents!" was neither righteous nor did it prove historically accurate. Yet true believers still try to excuse God's many erroneous predictions and broken promises (He even tried to kill Moses shortly after recruiting him, and was only deterred when Noah's wife Zipporah fended Him off with their son's bloody foreskin) by suggesting that God really meant it until those unruly Hebrews misbehaved.

But maybe God just forgot - in which case, Moses could have done the Israelites better service by having God carve His *Promises* in stone rather than His *Commandments.* Of course, it might be unfair to scapegoat God for His many reported deficiencies. After all, when not fighting each other, monotheists usually devoted considerable efforts to fighting the heathen. So if God really existed - and if He truly favored one group of monotheists over the others - then He certainly had innumerable opportunities to provide His Chosen People with a permanent military advantage - plus all the converts and recruits they might ever need - by leading them into battle just once with His Flaming Sword held high.

Yet despite countless truly desperate and deserving pleas for help, God never bothered - which actually makes sense. For ordinary logic assures us that the self-selected and obviously imperfect leaders of any religious sect are almost always better off representing a purely hypothetical deity than serving an actual judgmental one. Consider, for example, how badly God treated His Own Son who-was-perfect-in-every-way. But we are getting ahead of our story.

From Genesis to The Flood

Genesis was a Bronze Age effort to generate convincing history backward from a Middle East that was known to the mysterious beginning that wasn't. For monotheism would never persuade doubtful polytheists unless One God was shown firmly in charge from the start. After all, no objective new evidence had suddenly been unearthed that even slightly favored monotheism. Nor were there other obvious reasons for monotheism to be touted now, after millenia of successful polytheism.

Perhaps it was just Pharaoh's tantalising example. Or possibly another major environmental catastrophe finally emboldened the Hebrews to confront that entire polytheistic coterie of gods and goddesses. In either case, Pharaoh Akhenaten had shown the way.

How long ago is long enough?

By locating Creation almost three thousand years earlier, the Hebrew priests placed that beginning a few centuries before any known Middle Eastern historical event. This still left plenty of time to invent man's unique *In His Image* status - to explain *human knowledge* (that purloined apple from the Tree of Knowledge) - and to establish *man's God-given rule over other creatures.* Genesis also justified *woman's subordinate*, indeed, utilitarian *status* (a spare rib could never rule the whole), and it explained agricultural toil and painful childbirth *as divine punishments for Eve's transgressions* rather than serious imperfections in God's grand design.

The Flood Story simplified biblical genealogy by drowning all humans and animals except for Noah's family and two animals of every kind. And because the end of the Flood brought the Bible's tale toward historical times, that was an appropriate point to taper off those miraculous multi-century life spans in preparation for real human history. However, *an implausibly long life remained an easy way to flag important myths*.

So Abraham's impossibly long life and vigor represents an old story-teller's signal across the millennia that this tale is very important but not to be taken literally. The same rationale clarifies why Moses was a simple shepherd nearly eighty years old when God first called - identifying Himself as ***"I am that I am"*** - to make Moses an offer he couldn't refuse.

The Flood Story dealt with the mystery of those huge fossil bones found in the desert. It also eliminated any giants remaining on Earth from the early days (presumably those mysterious *nephilim* mentioned in the Hebrew Bible were well-known characters from an earlier fossil-based yarn). And the message we carry away is that God sees all, knows all, and can be devastating when pissed.

The Flood myth also reminds us that God (and His priests) are easily swayed by a good barbecue. Indeed, to reward Noah and his family for that great post-flood party, the storyteller tossed in the rainbow so that desert-dwelling Hebrew shepherds wouldn't rush to construct arks during every devastating downpour.*

*See also "Outcomes" on page 127.

Abraham

Abraham's story was an object lesson on the importance of hospitality to poor travelers - and what sensible traveler admits to being other than poor? For in the good old days before greedy ATM's - when it was still possible to find helpful human money lenders in the Temple - a hospitality requirement was especially relevant to otherwise exclusionary Hebrew monotheists who often traveled long distances to reach the Temple with their donations. After all, if the Hebrews despised and mistreated pagan travelers, they would surely be despised and mistreated in turn.

Consider the myth: God (cleverly disguised as a poor traveler - or was it a poor traveler cleverly disguised as God cleverly disguised as a poor traveler?) strolls across the desert with a couple of buddies. A simple old shepherd immediately sees through those clever disguises and throws a fantastic barbecue, pushing desert hospitality to its limit.

God so enjoys His favorite meal that He declares the old shepherd's future descendents His Chosen People forever. Furthermore, as a tip, God advises Abraham that his descendents can gain eternal rights to good land nearby by killing the current residents. God then gets Abraham's beautiful wife pregnant and the three travelers exit stage left, leaving the simple old shepherd scratching his head in disbelief at his great good fortune.

Moses

Well then, how about Moses and The Exodus? It is worth noting that Moses - *the only man who ever spoke "face to face" with God* - is widely revered as the most important person in the Hebrew Bible. For not only did God need to tell someone about His stellar works in Creation, He also chose Moses to lead those polytheistic Israelites to freedom and enforce His monotheistic Commandments.

Of course, God clearly exchanged some words with Abraham - and we may hope that He also made small talk with Abraham's lovely wife Sarah while getting her pregnant - and surely He grunted a few words to Jacob as they wrestled all night. But perhaps the woman didn't count. Or maybe God avoided being actually "face to face" with any of the characters involved in those more limited encounters.

Yet curiously enough, despite his primary role in forcing monotheism upon the Israelites and as leader of the Exodus, Moses was never once even mentioned by any contemporary historical account other than the Bible. It is especially difficult to explain why the Egyptians - after those devastating interactions with Moses - would have failed to put their own spin upon his miraculous plagues and the Exodus. For they certainly made careful notes about many less famous battles, hunts and encounters - even some that they lost. Hence a commonly held notion among Bible scholars is that the Exodus - as reported in the Bible - never happened.

Of course, it is possible that some semblance of these events actually took place during a great natural catastrophe which so overwhelmed the local Egyptians that they never had an opportunity to describe Exodus from the viewpoint of those left behind. Such an environmental disaster would also explain the plagues brought down by Moses as well as the ease with which departing Hebrews looted so many Egyptian dwellings of their valuables before the pursuing army from a less devastated part of Egypt caught up with them.

On the other hand, many scholars suspect that this entire tale was written - or extensively revised - many centuries later as enslaved Israelites made mud bricks during the Babylonian Captivity (597-538 B.C.E.). Among other reasons for their view is the knowledge that major Babylonian construction projects of that era featured mud bricks while Egyptian public works favored stonework.*

*See Moses, A Life by *Jonathan Kirsch.*
Also Moses The Egyptian by *Jan Assmann.*

As for Joshua's assault on Jericho soon after Moses' death, the priests and leaders could have proposed that campaign more truthfully - "Hey, guys. You know what? This land becomes our land, at least for a while, if we can take it by force". But then the Hebrews might not have fought as fiercely or killed all the locals. And those displaced survivors could have rebuilt their forces and counterattacked.

Hence it seemed more logical and effective - although morally reprehensible - to simply murder the entire population of Jericho and blame God for absolutely insisting upon all of that terrible carnage. Here we have another early example of religion being used to justify "ethnic cleansing". And we again meet that age-old excuse "I was just following orders."

The Bible tells us that the priestly writers of Exodus - probably then living in Babylon - had difficulty motivating recently freed Hebrew slaves to regain their ancestral lands and restore the Temple. For it seems that many of those ex-slaves rather enjoyed their new-found freedoms and life in the big city. So why should they return to the desert and restore the Temple when that would simply encourage the priests to further dominate and tax them?

Therefore, to encourage their return, the Bible exaggerated Abraham's close relationship with God and with the Promised

Land - based entirely upon Abraham's irrational faith and God's obvious enjoyment of that great barbecue and other hospitalities. But while Abraham's actual existence as a historical rather than mythical character remains an unsettled issue amongst archeologists - the whole barbecue outcome might simply represent a misunderstanding.

For example, suppose God sent His usual compliments to the chef, and poor old Abraham took this to mean that he might already be a big winner. Maybe God said He'd try to get back for another cookout in years to come, and Abraham assumed this meant that his descendents would be God's Chosen cooks forever. In any case, the priests surely found it difficult to convince the Israelites - newly freed from two generations of cruel slavery - that they really were God's (or anyone else's) Chosen People.

So when subtle priestly propaganda and Biblical tall tales proved insufficient, the priests finally just commanded the Hebrews to go home and resume their ancestral ways as nomads in tents - insisting that they give up the soft life - *"Just say No!" to coveting the neighbor's house - Tenth Commandment* (might Moses have received only Nine Commandments during the Exodus?) - and leave behind the many other temptations of beautiful downtown Babylon, including those world-famous hanging gardens.

Chapter Five

****

"Your honor asked me whether evolution has anything to do with the principle of the virgin birth. Yes, because this principle of evolution disputes the miracles, there is no place for the miracles in this train of evolution, and the Old Testament and the New are filled with miracles. If this doctrine is true this logic eliminates every mystery in the Old Testament and the New and eliminates everything supernatural, and that means they eliminate the virgin birth - that means they eliminate the resurrection of the body - that means they eliminate the doctrine of atonement and that they believe man has been rising all the time, that man never fell, that when the Savior came there was not any reason for His coming; there was no reason why He should not go as soon as he could, that He was born of Joseph or some other correspondent and that He lies in His grave; and when the Christians of this state have tied their hands and said, 'We will not take advantage of our power to teach religion by teachers paid by us,' these people come from the outside of the state and force upon the people of this state and upon the children of the taxpayers of this state a doctrine that refutes not only their belief in God but their belief in a Savior and belief in heaven and takes from them every moral standard that the Bible gives us.

"...I have been so well satisfied with the Christian religion that I have spent no time trying to find arguments against it. I have all the information I want to live by and die by."

William Jennings Bryan (at the Scopes trial, 1925)

****

Christian doctrine was a very complex revision of Hebrew Monotheism's complex revision of Egyptian Monotheism

The background to Christianity - a brief review:

Monotheism originated as a simple challenge to Egyptian polytheists: "Almighty God is The Universal Creator. The Eternal One. The Only True God. The Sun in our Heavens. He rules over all. And I, Pharaoh Akhenaten, His Only Son, will henceforth lay His orders upon you and accept your donations for Him." The One True Sun God was thereafter worshipped for about 20 years until His Only Son died naturally - at which point the people shrugged and went back to polytheism.

An unknown number of years later, the One True Eternal God was suddenly resurrected, revised and returned to active duty by Hebrew priests who carefully provided Him with a totally new, purposely nondescript and non-solar identity: ***"I Am That I Am"***. It turned out that this enigmatic self-description - like *the sound of one hand clapping* - was practically perfect for pondering on an otherwise slow day.

In addition, the Hebrew priests fabricated a brief history of the universe that explained Creation and revealed how Abraham's generous hospitality - plus his deranged willingness to put blind faith ahead of reason and sacrifice his own son to the one True God (might this reflect a delayed onset of doubt about the child's paternity?) - so impressed Almighty God that He made the future Israelites His Chosen People forever (though a good foster home for young Isaac ought to have been His first priority).

The priests further disclosed that the Hebrew "Chosen" status included eternal entitlement to valuable real estate currently occupied by others - and this knowledge was important, for men and animals fight far more fiercely to defend or regain what is rightfully theirs than to pilfer something to which they have no obvious claim. And the priests reminded the

Chosen People that all bribes addressed to God should be dropped off at the Temple for forwarding.

The Israelite model of monotheism took root and persisted because it gave a loose-knit and unruly Semitic tribe a wealthy centralized organization and common purpose, which in turn lent *them disproportionate military clout.* Of course, the hereditary priests had a vested interest in sustaining monotheism and their own ample tax-free incomes. But even ordinary Hebrews appreciated God's gracious support for their otherwise groundless claims to good land.

Yet that land may also explain why Hebrew monotheism remained small and exclusive. For a faith based upon ethnic preference and limited real estate could not readily recruit outsiders without reducing each shepherd's share. So while militant monotheism underlay Hebrew military successes, their God-given claim to The Promised Land kept Israelite armies from expanding to become more than a regionally credible military force.

On the other hand, Christian and Muslim theorists carefully avoided placing similar ethnic or geographical restrictions upon their own power, hence the bloody conflict between Christianity and Islam eventually extended around the globe. This apparently irresistible spread of monotheism was aided by the devastating epidemics that reliably accompanied urban invaders into new territories, as well as by technological advances and the domestication of animals.*

*See also Guns, Germs and Steel by *Jared Diamond*

In any case, monotheism's insistent centralization of wealth and power - plus its ability to incite and sustain religious fanaticism - usually gave Christian and Muslim soldiers a clear military advantage over the heathen. Yet their world-wide deployment, similar power structures and comparable levels of fanaticism and technology kept either Christianity or Islam from finally obliterating the other through battle or by conversion.

That global stalemate between wealthy warring monotheistic faiths - along with equally ferocious infighting amongst their various factions and sects - has caused endless human misery, death and destruction over the past thirteen centuries. Nonetheless, loyal long-suffering members of each group continue to pray for God's help in destroying the others. And somehow, many Christians and Muslims still manage to interpret their ongoing military standoff and unbridgeable doctrinal differences as proof that God wants the killing to continue until their side has won or no one is left.

Christian and Islamic doctrines are ultimately founded upon the same Hebrew Bible tales. Yet their reliance upon the Hebrew Holy Book seems unjustified, given God's obvious failure to reward devout Hebrews with any outstanding military victories for several thousand years. After all, one cannot logically credit more recent spectacular military successes by *secular* Israeli soldiers against devout Christian and Muslim Arabs to either the Hebrew Holy Book or to favoritism by its divine author.

The Birth of Christianity

Christianity originated during a period of intense social upheaval. Frequent military invasions and local revolts against distant rulers repeatedly disrupted and mingled previously distinct Middle Eastern populations. The Temple-based social order was in total collapse after wealthy religious leaders absconded or were carried off with their treasures. Nonetheless, religious belief remained strong and various anthropomorphic gods shared credit or blame for whatever transpired.

In an age when magic and miracles were ordinary aspects of daily life, many disoriented individuals from disrupted groups sought stability within associations of like-minded individuals. Naturally they were eager to fit in and learn from their new friends and neighbors. Not surprisingly, the identifying facts about an individual's skills, origins and ancestry were presented in the best possible light. And almost everyone had access to some ancestral myth featuring an appropriately embellished line of descent from the gods. Story tellers and poets were held in high esteem. Old tales improved with each telling.

The Jesus sect began as a landless group of itinerant preachers who followed a Jew named Jesus. From the start it was an inclusive organization, drawing the poor and dispossessed from all groups. With little or nothing to share, these early members depended upon charity for sustenance and their charismatic leader for inspiration. But after the natural death of Jesus, leadership squabbles soon disrupted his sect into several feuding factions. Those in the Christ faction had a special talent for composing miraculous tales about Jesus that brought additional recruits to their group.

Christian doctrine is impossibly difficult because it had an impossibly difficult problem to solve

Christianity was written and produced by the Christ faction (christ is a Greek term for "the anointed") of those who were spreading the teachings of Jesus. But while many Jews met terrible deaths under Roman rule, the crucifixion tales about a Jewish sage named Jesus first surfaced several decades after that supposed event. Other New Testament myths about Jesus and his miracles originated even later, as subsequent generations revised and enhanced the Christian story.*

*See The Lost Gospel *by Burton Mack.*

Many of the miracles eventually attributed to Jesus seemed familiar and believable because they already were prevalent in Middle Eastern myths. These included predictions of his coming - a virgin birth - walking on water - healing by touch (long practiced by many kings) - restoring sight - feeding the masses - raising the dead - reincarnation. In fact, these particular marvels glorified the lives of quite a few major prophets and important rulers in that time.*

*See The Bible As It Was *by James Kugel.*

The New Testament

The Hebrew Bible was a document categorically constructed for the Jewish core constituency to which Jesus belonged. Hence Jesus was already comfortable with its mysterious God, and often spoke of Him in his own teachings. But after Jesus's natural death, the ongoing failure of Christianity to attract a strong Jewish following left Christian legitimacy awkwardly balanced upon the Greek translation of a Hebrew Holy Book. And this soon led to increasing doctrinal difficulties.

Thus the New Testament project became a high priority for the Christ faction of the Jesus sect. Naturally, the New Testament editors developed as many parallels as possible between their tales about a relatively unknown Jesus and the beloved Hebrew Bible stories about Israel's revered ancient leader, Moses. Hence even though the New Testament was written to establish Christian credibility, it had the unfortunate side-effect of reemphasizing how much Christianity and its New Testament depended upon analogies between Old and New Testament stories, and on idiosyncratic Christian interpretations of Hebrew Bible teachings and prophecies.

For example, the early Christians who reissued the Hebrew Bible as their own Old Testament, managed to modify the sequence of Hebrew Bible stories and prophecies sufficiently so that a several-centuries-earlier peak in Hebrew Messianic fervor would seem to reach its climax just before the "long awaited" advent of Jesus. Naturally, this dramatic juxtaposition made it even harder for later generations of Christians to understand Hebrew disinterest in Jesus as the Jewish Messiah.

Early Christianity had an anti-family bias

Although they often were awkward, and some sound downright weird to the modern ear, the teachings of Jesus and his followers - as well as their itinerant lifestyle - made early Christianity a naturally anti-family organization. Consequently, when someone said to him, "I will follow you wherever you go,"

Jesus answered "Foxes have dens, and birds of the sky have nests, but the son of man has nowhere to lay his head." When another said, "Let me first go and bury my father," Jesus said, "Leave the dead to bury their dead." Yet another said, "I will follow you, sir, but first let me say goodby to my family." Jesus said to him, "No one who puts his hand to the plow and then looks back is fit for the kingdom of God."

And almost a century later, in Revelations, John speaks of the *"one hundred and forty-four thousand who have been redeemed from the world. These are the ones who have kept their virginity and not been defiled with women...(they will serve as) first fruits for God".* A similar homoerotic, elitist, anti-family bias may underlie the modern Church's insistence upon priestly celibacy, as well as its grudging attitude toward women and heterosexual intercourse.

That bias could also explain the widely presumed prevalence of homosexuals - and the alleged high incidence of AIDS, as well as the remarkable tolerance for pederasts - within the present-day priesthood. Comparable tendencies and behaviors are reportedly still common among desert dwelling Arabs and other Muslim males.*

*See, for example, Caravans by *James Michener*, or The Seven Pillars of Wisdom by *T.E. Lawrence.*

On the other hand, their anti-family stance saved early Church Fathers from organizational difficulties of the sort that confronted hereditary Hebrew priests (too many entitled descendents - all from the same ethnic group), which might have endangered the Christian Church's future expansion and patronage. And with no job-hunting descendents to look out for, the itinerant Church "fathers" had no basis for an inherited priesthood anyhow.

The Church they designed is therefore run by an eternal self-appointed old-boy network - led by a pope elected by cardinals who were appointed by other popes who were elected by other cardinals in an infinite regress. And once that tight leadership was in place, all Christians were encouraged to worship graven images of the One True God, or of *Jesus on the Cross,* or to pray before various depictions of the Blessed Virgin or before icons or even fragments of various saints, since there was no longer any significant risk that Church income might thereby be diverted.

Not surprisingly, the fable of *Jesus on the Cross* required regular reinforcement to become and remain established truth. Worshipping *Jesus on the Cross* was also an appropriate way for the Christian Church to emphasize his importance. For just as Christians planned to supplant the Hebrews as God's Chosen People, adoration of *Jesus on the Cross* openly flouted Hebrew Bible strictures and God's Second Commandment about idols and graven images. Clearly, the major religious transformation being made in Jesus's name more than justified all of these new rules.

Naturally, not every prelate favored each of these doctrinal "advances" as they were voted in. So to maintain proper spiritual harmony among the leadership, it became customary to persecute those who had voted with the minority, especially if - rather than promptly recanting their erroneous views - they then had the temerity to demand a recount or to continue promoting the minority position. For public disagreements about church doctrine were viewed as *heresy, sacrilege or blasphemy* - all capital crimes - since the unquestioning unanimous support of every "approved" Church position was what sustained Christian credibility, wealth and power.

Hence the Church always responded quickly to anyone who might challenge its self-appointed role as the judgemental controller of a postulated pipeline that transported hypothetical souls to a theoretical Heaven. *Indeed, the Church's greatest fear was that some credible observer might publicly suggest* that the Emperor wore no clothes - *that the pope's theology was a transparent fraud - and get away with it!* After all Monotheism's public challenge to polytheistic gods and goddesses had been its defining moment. And the failure of those deities and their human handlers to respond viciously had opened the way for a new belief in One God.

Yet the process of subordinating and eventually enslaving all Christians to the Church really started when Christians first utilized the mythical forms of praise popular at that time to divert attention from Jesus's ordinary death. Indeed, the Church repeatedly revised and replayed that demise in New Testament Gospels and elsewhere as the crucifixion myth gradually evolved into the central Christian drama, replete with heroics, virtue, doubt, despair, betrayal, resurrection and divinity. This imaginary event even made possible the later determination that (like Pharaoh Akhenaten before him) Jesus was the Only Son of God - until that idea was outgrown as well.

However, Christianity's big break finally came early in the fourth century when the leadership managed to convince Constantine - the nervous new emperor of an increasingly threatened Roman Empire - that he could ensure his wealth and power by requiring all Romans to embrace Christianity as the new state religion. Not long thereafter, Church fathers realized that they no longer had to explain the complex and confusing theology of Christianity, nor try to make it more convincing. And from that time forward, they simply stressed the importance of faith in miracles and mysteries, while the swords of Christian soldiers spread the word about the Prince of Peace.

Chapter Six

The camel is a horse designed by a committee

It has been said that the camel is a horse designed by a committee. For the domestic camel is a rude and ungainly beast whose various parts just don't seem to fit. In fact, the female lies down when sexually approached, so it even requires assistance from its human handlers in order to mate with a male camel. Presumably this tendency to lie down inappropriately is an inadvertent side-effect of selectively breeding those camels that were most useful as beasts of burden, since a standing man cannot conveniently load a standing camel.

Similarly, of all the great monotheistic faiths, Christianity most clearly displays an awkward consensus and the many inadvertent side-effects that resulted from numerous committee decisions made over several centuries. Indeed, many generations of early Christians had to endure tortuous theological tumult as one committee after another revised religious history and rewrote Church dogma in their unending effort to show that the Christians had finally replaced the Hebrews as God's Chosen People.

From the start, the Christ faction appeared determined to create an entirely new theology *loosely* based upon the life of the group's late lamented leader. Of necessity, that creative process only reached full fruition after many generations of believers - who had never heard Jesus speak - worked their way up the organizational ladder by issuing fervent and original praises of him. And this process still goes on.*

*See, for example, " a bold new image of Jesus " on page 102.

So what were the major stumbling blocks that interfered with the spread of Christian belief? And how were these overcome or circumvented?

Well, early Christian theorists certainly had many difficult facts to deal with. For example:

Jesus was a Jew, as were some of his original followers.

Many followers took Jesus to be the long-awaited Hebrew Messiah.

Yet few Jews accepted that claim (hence those "perfidious Jews" were disparaged during every Good Friday prayer until the 1960's).

In addition:

Jesus appears to have died an unremarkable (natural) death, and soon Christianity became a non-Hebrew faith based upon the Hebrew Bible.

But the Hebrew Bible insisted that the Jews were God's Chosen People.

And to dispute that Bible would undermine Christian legitimacy.

Therefore, Christianity needed a uniquely powerful message. So:

***Christians invented an eternal soul** that God could reward or punish.*

*They devised **heaven** and **hell** as venues for rewards and punishments (or to be more exact, they customized the sixth century BCE Persian Zoroastrian belief that divine justice would be meted out in the afterlife)**

* See Armageddon in the Bible *by Gerald A. Larue in "The Humanist" November/December 1999*

They invented Original Sin - a depravity inherited from Adam and Eve.

*They decided that God was still furious about the(se) **Original Sin**(s).*

They invented Jesus's crucifixion and made it the central Christian drama.

*They confirmed his crucifixion by always depicting **Jesus on the Cross.***

They blamed his ***crucifixion*** *on the Jews - though God allegedly planned - and the Romans supposedly carried out - that hypothetical dirty deed which has provided so many real priests with comfortable lifetime jobs.*

And once Jesus's death had somehow atoned for humanity's sins, eternal souls could finally be salvaged through theoretical visas issued by the bureaucrats of Jesus's church to invisible souls for entry into hypothetical heaven.

But even this didn't suffice, so:

They made Jesus son of man into Jesus Son of God, then Jesus part-of-God.

*They made Jesus's mother Mary into a married virgin so God could get her pregnant. Divine adultery was fine, but not out of wedlock, since other single mothers might then claim "a virgin birth!" to avoid being stoned to death.**

*See also "Unwed Fathers" on page 144.

But the Hebrews still insisted THEY were God's Chosen People!

So to prove that Christians REALLY were better than perfidious Jews:

The Christians disrespected Hebrew dietary laws.

They worshipped Jesus on Cross and other idols.

They persecuted the Jews for not accepting Jesus.

They persecuted the Jews for encouraging Romans to arrest Jesus.

But they couldn't criticise the Romans for killing Jesus since Romans kept good records (and might eventually become Christians).

And finally, in desperation, they wrote Revelations to show that on Judgement Day, all those who had rejected Jesus would be Unchosen and greatly punished. To summarize that outcome in modern terms, *God has prepared* ***eternal rewards*** *for the faithful who ignore or deny the scientific findings that make our technological civilization possible. And God has prepared*

eternal tortures *for all the sensible folk who make their choices based upon confirmed evidence and ignore totally unsubstantiated (creationist, or other) claims.*

Indeed, Christian fundamentalists happily interpret John's feverish visions reported in Revelations - not as a ghastly nightmare brought about by some major dietary indiscretion, serious mental illness or temporal lobe epilepsy, but rather as proof that on Judgement Day, God intends to carry out ethnic cleansing on a scale and with a ferocity that should make Hitler (that soldier of Christ) appear kindly in comparison.

Furthermore, they expect God to instigate this incredibly cruel mayhem while allowing 144,000 good Christian youths "who have kept their virginity and not been defiled with women" to follow Him around so these "first fruits for God" may be available to satisfy His weird whims while God devises eternal torments for ordinary heterosexual Christians and all the others who ignored Christianity's implausible message and deranged threats.

In the meanwhile, based upon the Eden/Apple story and the obvious inadvisability of disagreeing with, or even questioning, such a mean-spirited God, Christian Fundamentalists plan to remain safely ignorant of everything not explained or hinted at in the Bible. Of course, ignorance encourages poverty, and many Christians view wealth as a sign of God's favor. Consequently, many poor creationists find the tax-free wealth flaunted by their religious leaders reassuring, even though those leaders extorted their riches from ignorant fundamentalists by implausible threats and promises.

The Christian Doctrinal Dilemma

Christian dependence upon the Hebrew Bible (which identifies the Hebrews as God's Chosen People) ***made Christianity hard to market*** since it could not logically offer the inside track to God that every religious convert had a right to expect. Thus Christian theology is best understood as a mechanism for filching the Chosen status from the Jews through a circuitous sequence of ad hoc revisions and absurdly leveraged amendments. But ultimately that theft failed to make sense and it left Christianity looking *logically* unattractive. So the Christians decided to degrade, despise and oppress all Jews in order to make them look *emotionally* unattractive (hateful).

At the same time, *Christianity continued to grow* primarily *through military conquests* (as when the pope split South America between Portugese and Spanish Catholics) *and by its totalitarian control over pregnancy and abortion* (the latter usually being considered a serious criminal act that denied an innocent soul the opportunity to become a soldier of Christ and perchance reach heaven). Thus Christianity grew through coercion and reproduction rather than via thoughtful conversions.

"Give me a child before he is five and I will give you a Good Catholic"

Like other irrational prejudices, religion is most firmly rooted if ingested with mother's milk. Indeed, it early became apparent that religious myths taught to young children as they developed their language/logic faculties were rarely rejected later in life, *for language development precedes and underlies rational thought.* Therefore, when those *brain-washed children* entered adulthood, they found it impossible to rationally discuss or defend such deeply ingrained religious illogicalities.

In fact, were they able to verbalize them in a coherent fashion, many religious persons would undoubtedly inspect, be appalled, and forthwith abandon their irrational beliefs. But they cannot, so theological arguments are famous for producing more heat than light. And religion remains a profitable - therefore, self-sustaining - belief system that encourages social solidarity at the expense of outsiders.

That *pre-logical implantation of religiosity* also underlies the notoriously heterogeneous beliefs of ordinary Christians. Even members of the same congregation, reading from the same prayer book, soon recognize that their personal religious truths differ markedly from those of their neighbors.

Interestingly, the acquisition of religion both coincides with language acquisition and resembles it. So just as certain sounds peculiar to a particular language may remain undetectable and unpronounced by persons who learn that very different language later in life, those of us with no early immersion in religious thought often find that we have a tin ear for religious input in our later years. We simply don't get it. It just doesn't compute.

Jesus died for your sins

Of course, Jesus' allegedly sacrificial death didn't exactly elicit God's final forgiveness for your particular personal sins - whether major or minor. After all, even Christ's Magical Kingdom cannot claim that the *effect* (His alleged crucifixion) preceded its *cause* (your alleged sins) by two millenia. But the misleading marketing mantra headlined above will have done its job if it elicited even a tiny involuntary pang of guilt - some remembrance of a less-than-stellar moment - a fleeting feeling of relief and forgiveness, and especially - a sense of obligation to the Christian Church.

Nonetheless, one might hope that if God did prearrange Christ's crucifixion, at least Christ suffered to atone for some truly heinous human failing. So exactly what sins do all humans share that necessitated Jesus's sacrifice? Unfortunately, no one seems quite sure. Or if they know, they aren't telling. Thus we marvel: Was God still sore over losing *one apple?* How could He possibly remain enraged about such a remote and natural transgression after the innumerable truly loathsome and premeditated evils that man **in His Image** has committed **in His Name** ever since?

A deity never forgets

Any objective observer would surely agree that the abrupt expulsion from Eden - and God's decision to invent weeds and institute painful childbirth (along with stretch marks, hemorrhoids and other varicosities) - and His destruction of almost all life with that great Flood - ought to have soothed even the Almightiest Annoyance. And surely any mortal would deserve swift censure and institutionalization or possibly capital punishment for seeking out a married virgin to impregnate so that he might later kill their son to compensate himself for an apple eaten thousands of years earlier by some harmless primitive couple.

So was God really upset about that act of disobedience? Or about the sexual experimentation that followed? In either case, the last thing an intelligent all-powerful God ought to want would be for a church to organize itself as an ongoing celebration of His criminality and sorry betrayal of marital vows - not to mention the many other gross violations of His own Commandments. But perhaps such concerns were beside the point. For the priests redesigning Christian doctrine faced a formidable problem.

They had to come up with an additional emotional draw, and do it soon. They needed a new supra-tribal rationale that could attract more recruits to their struggling, unintentionally non-profit, organization. Yet everyone who might even consider becoming a Christian faced the same dilemma. Why join a monotheistic faith that the Israelites themselves were ignoring when this new faith derived most of its legitimacy from the Hebrew Bible - and especially when that Bible clearly identified the Israelites as God's Only Chosen People?

How could anyone be recruited with such stirring words as "Follow Christ and become one of God's Unchosen!"? What religion could prosper by honestly admitting that God ignored all prayers by its dues-paying members. Yet the priests already had duct-taped every available miracle to Jesus' life story. They already had depicted Jesus as far less radical than he actually was. And still Christian missionaries were mostly met with a big yawn.

So a majority of the prelates voted that *Jesus of-the-virgin-birth* was actually *the One and Only Son of God, even though he-who-always-spoke-the-truth had repeatedly declared himself "the son of man"* - which surely made more sense. But after the Christian Church worshipped the One True God and His Only Son for a while, they decided that His Son had actually been part of God Himself all along - since this was more in keeping with monotheism and God's undivided omnipotence.

Besides, God-in-Christ meant God really hadn't killed His Only Son to compensate Himself for one stolen apple, but rather that God had actually provided a part of Himself for crucifixion and resurrection to show how much He appreciated those who would later be born - or coerced into becoming - Christians.*

*See also "The Holy Trinity uncovered" on page 139.

Chapter Seven

"Eternity" was the key Christian innovation

Christianity's convoluted doctrine was constructed in increments. Most of those sequentially added parts represented ultimately unsuccessful efforts by church authorities to plausibly position Christians as "God's Chosen People". For regardless of whether Jesus was eventually defined as a great prophet, or as the Son of God, or as an actual Chunk-of-God, Christianity would live or die on how many new followers it attracted and how completely it controlled them - so it urgently needed a way to reward the faithful and punish back-sliders.

However, no Church dedicated to the centralization of wealth and power can grow by passing out expensive rewards. Nor - until it achieves totalitarian control - can it risk alienating believers by dispensing many major punishments on Earth. Therefore, the Christian writers devised *eternal rewards* and *eternal punishments* applied to *eternal souls* - none of which the Hebrew Bible had even mentioned.*

*See also "Pascal's Formula" on page 114.

Furthermore, Christianity invented *Heaven* and *Hell* - as well as *Limbo*, the postulated warehouse where unbaptised good souls were stored until Pope John Paul II recently declared Limbo an unprofitable fabrication that the Church would clear out and discontinue. And, at least to date, Christianity remains heavily invested in *Purgatory* - as described by that great Italian poet and fiction writer, Dante (1265-1321).

In fact, once Purgatory was officially affirmed as an independent profit center in 1563, it became the mandatory pit-stop wherein souls not yet qualified for Heaven could suffer for an interlude that might be modified by appropriate prayers and payments. The Church also discovered *Original Sin* - mankind's awful inherited crime - which neither the Hebrew Bible nor Jesus had ever mentioned. And at that point, this Rube Goldberg mechanism for removing money from the pockets of the poor, the ignorant, and the defeated, ran approximately as follows:

Jesus died for your sins (some unspecified, some **original**), **so your** hypothetical (invisible, only recently defined) **eternal soul finally has an opportunity for redemption through the Church** that Jesus might have invented had He thought of it before His ordinary death. For most of you, that redemption will mean *poverty* (so give the church what little you have), *chastity* (but if you must indulge in sinful heterosexual behavior, at least make as many children as possible in order to expand the supply of desperately poor workers, soldiers and sailors) and *obedience* (or we shall remit your eternal soul to a very bad place where it/you will suffer terribly *forever!).*

On the power and irrationality of Christian Guilt

Christian guilt rapidly became an awesome force for social discipline as the ignorant masses finally accepted the Church's claim that life was not a random, nasty, brutish and short event as it appeared - but rather an elaborate hoax - a divinely disguised trial in which the most intimate behavioral detail - the loss of virginity, perhaps - or masturbation - or momentary disbelief - might critically tilt the final determination on whether a soul deserved eternal bliss or was earmarked for eternal torment.

Of course, Christian theologians never tried to explain how a trial so plagued by deceptive appearances, misleading clues, unbelievable dogma, purchased absolutions, invisible witnesses, postulated judges and hypothetical, nearly random, risks and rewards, could be anything but a crooked charade. Nor would they provide an objective basis for estimating how many points one still might need in order to overturn or reverse an allegedly adverse divine decision, as even that information might loosen the arbitrary controls by which the Church impoverished and enslaved all Christians.

Furthermore, no one ever brought up God's immense investment in this inconceivably huge and complex universe with its countless galaxies and stars; nor did anyone try to explain how an entire obviously ancient (though newly created) universe could logically be dedicated to the testing of invisible souls on one minor planet. Even more amazing was the idea that God so enjoyed rewarding good souls and punishing the less-than-perfect ones that He Himself had just manufactured, that He would devote an eternity to this pointless task. And perhaps most amazing of all was the idea that He might then be sufficiently concerned about minor items such as the profitability of His Church, to permit the sale of absolutions for sins - and thereby bypass His reason for creating that entire universe in the first place.

A strong Church doesn't reason, it demands

A) All Christians will pray to the God we have postulated. Or else!

B) All Christians will support His self-declared representatives in proper style even if donors then do without.

C) All Christians will accept the unlikely notion that God has manufactured and installed billions of undetectable entities known as souls.

D) All Christians will agree that each of these invisible souls is far more important than anything that exists or could occur in the real world.*

*See A World Lit Only by Fire by *William Manchester.*

Furthermore;

E) All Christians shall dread being misled by God's (equally hypothetical, similarly invisible) immortal enemy known as *Satan,* whom

F) God kindly appointed to oversee Hell - the invisible domain wherein ghastly eternal punishments are applied to invisible souls that fail to pass invisible God's alleged inspection while

G) God Himself implausibly oversees a lavish musical section that allows Him to eternally reward invisible souls that were unfairly judged to be perfect.*

*See also "Abortion and the Human Soul" on page 105.

Of course, it was never made clear how one illogical and hateful hypothesis - the Christian God (one of whose clever redesigns - reported in Revelations - was hordes of locusts that stung like scorpions and caused such great pain for five months that men wished they were dead) could force another purely evil hypothesis (the Christian Devil, over whom God had no apparent control) - to reliably deliver God's unjust punishments.

Indeed, one might logically expect the Devil to run a pleasant two-step ice cream parlor/coffee shop program for sinners, and to otherwise interfere with God's planned punishments wherever possible. It therefore seems probable that any authentic groans emanating from Hell - appropriately amplified for God's edification - are elicited by ordinary ice cream headaches. So anyone truly concerned for their own hypothetical soul might want to pray that it goes directly to Hell, which apparently is the only way a soul can escape God's illogically punitive oversight .

We need not ask what the Church gained from all these ill-founded hypotheses, as the wealth and power of the Christian Church are there for all to see. But the question Christianity has always avoided is "What is in it for God?" Why the hell would any ordinary mind - let alone a postulated infinite intelligence - ever embark upon such a silly project, the likes of which would soon bore the average three year old. But to the faithful, that simply doesn't matter. "For God works in mysterious ways". "Faith will be rewarded". "It is all for the best". And so on.

Salvation

Thus guilt and self-inflicted punishment spread widely, driven by the fearful idea that even minor sins might prevent the salvation of one's eternal soul. Indeed, the persistent strength of Christianity arises ever again from the life-long preoccupation of each true believer with his or her own personal salvation. And where every act endangers one's ***eternal*** soul, who can afford to view others as aught but a stepping stone toward personal salvation?

Different individuals have responded in wildly different ways to the ***Christian emphasis upon personal salvation through absolute selfishness.*** Some initiated the Spanish Inquisition; others were drawn to the destructive and greedy invasion of the Americas. Yet on the more positive side, we might include the scientific explorations of Magellan, as well as some inner-city religion-supported educational institutions and hospitals. And we ought not forget the sterling social work of individuals such as Mother Theresa - though even her best deeds were inevitably tarnished by Christian righteousness and a presumption of moral superiority.

Still nobody can deny that Christian belief has stimulated marvelous art, or that certain monasteries produce superb wines, liqueurs or cheeses. And surely an occasional wealthy Christian has tried to help the poor for their own sake. Nonetheless, it has long been understood that wealth - however acquired - was more likely to retain God's blessing if shared with the Church.

Furthermore, the conservative rich and the Church leaders (often nearly one and the same) have historically viewed ordinary mortals living in abject poverty as unlikely candidates for salvation. So the poor probably deserved little assistance in this world anyhow, except as incidental beneficiaries of good works that would draw God's favorable attention to more wealthy and deserving Christians.

On the other hand, punishments remained cheap and readily available. They also provided low-tech fun for ignorant believers, with essential supplies as near to hand as the closest wood pile. Hence while the virtuous had to wait patiently for their Heavenly rewards, negative rewards were applied unstintingly right here on Earth so that all might get a feeling for the eternal torment that so many were promised.*

*A World Lit Only by Fire by *William Manchester.*

How Christian Belief is Confirmed

We have seen that internationally recognized polytheistic deities were known by different names in different cultures. And since every foreign name indicated broader acceptance, each lent its god or goddess additional clout. By the same token, the One True God gained further credibility as He became the focus of three major religions. So while the Hebrew Bible eventually supported and validated three separate monotheistic faiths, each of these persuasions in turn reconfirmed the Hebrew Bible.

Interestingly enough, this sort of subjective and circular self-substantiation is all that underlies the credibility of any religion. Hence what true believers really crave - but never can find - is *any* objective evidence that *even slightly* favors their particular religious hypothesis - which ought to tell them something.

But rather than interpret that absolute absence of confirmatory evidence as a sign of something terribly wrong with their beliefs, believers flock to worship anything from the presumed actual toe bone of a saint, to the scientifically discredited Shroud of Turin, to the site where imaginative girls sighted the Blessed Virgin, to the photo of a cloud resembling Jesus.*

*See also "Every Sunday We Pray" on page 130.

Still the question arises whether an objective way might exist to validate or invalidate various religious beliefs and procedures. For example, there ought to be detectable differences between a world in which some prayers are answered and a world in which none are. Let us begin simply by evaluating ***public prayers.*** Quite ***clearly,*** these ***don't work.*** At least, they don't reliably stop wars, win ball games, or turn rain "on" or "off".

Well, what if private prayers were answered just occasionally? To keep it simple, let us postulate that *God answers only the most urgent pleas of the very best Christians.*

Under such circumstances, the world would be plagued by miracles. There would be no predictable cause and effect. Science would be impossible. The best Christians would remain eternally healthy, wealthy, young, beautiful and wise. So we can easily rule that out.

On the other hand, what if private prayers all had equal weight - and many, most or even all were granted. Such a world might initially seem unchanged since it could only respond statistically to the preponderance of prayers on every subject. But before long we would correctly begin to blame each other for things that went wrong, or attack those who probably wished bad luck upon us, or redouble our own prayers for evil to befall others before they did us in. Thus *all prayers being answered* would soon throw the world into a downward spiral that none might survive.*

*See also "Heaven[ly?]" on page 123.

In fact, the obviously catastrophic potential of frequent divine interventions makes the random rigors of our naturally selected world seem almost attractive. So despite an occasional win, cure, survivor, death or disaster that defies all expectations, the case for or against *any* divine intervention remains unproven at best - especially as one might logically expect a truly divine response to be overwhelmingly apparent - as in "Oh No!!! Not The Flaming Sword!!!"

Yet if we back away from the confusing details of everyday existence and try to view the big picture, we still might *test the ultimate truth of any successful religion by first studying it in every detail, and then postulating an almost identical religion that differed only in having an* ***admittedly hypothetical deity*** (D). Then, provided there is no detectable difference between religion (X) as it now stands, and religion (X) with only a hypothetical deity (X minus D), one would have to conclude that **X=X-D**. And in this case, the simplest, most obvious explanation has to be that **D=0**.

Or to reason it out in words:

If a thing works equally well with or without some postulated, critically important, invisible part, then it just adds unnecessary complexity to postulate that critically important invisible part.*

*See also "Kindly Skip the Funeral" on page 118.

Chapter Eight

****

Theistic evolution is "an anaesthetic that deadens the Christian's pain while his religion is being removed".

William Jennings Bryan (1925)

****

Religious faith confirms dogma -

Religious dogma confirms faith

In the end, the credibility of any hypothesis or belief rests upon the standard of proof it must meet - and that standard may change significantly between noon in a science lab and midnight in a possibly haunted house - or even after a few drinks. But at least in an American Court of Law, sober jurors are routinely called during the day to determine whether a "preponderance of the evidence" supports a certain claim, or whether "clear and convincing evidence" reveals that an alleged event actually took place, or whether guilt has been established "beyond a reasonable doubt".

In contrast, it is widely held that *faith is a gift from God*, since only those with faith believe religious dogma and only those who accept religious dogma have faith. Could there possibly be a better example of circular reasoning? Well, perhaps.

It turns out that similarly self-substantiating evidentiary standards support the "New Rite of Exorcism" recently released by the Vatican in a Latin version and handsome red leather binding. For the Catholic Church still firmly believes in the Devil and his ability to lead the unwary astray. Indeed, from their point of view, *the fact that so many people think the Devil is non-existent, simply confirms what a clever devil he really is!*

One gathers that those who are intent upon believing must be willing to grasp at straws, disregard or reinterpret all adverse evidence and return regularly to replenish their faith at life's bottomless well of anecdotes and random coincidences. But just as the end of the Cold War caused consternation and consolidation throughout the military-industrial complex, so the remarkable recent decline in nefarious activities by the Devil has aroused Vatican concern.

Of course, there are those who blame increasingly effective psychiatric and epilepsy medications for ruining the Church's formerly bustling business in exorcisms. Yet this merely reemphasizes how the inexorable growth of scientific knowledge continues to shrink the inexplicable - and how the recent dearth of mysterious outcomes that one might reasonably attribute to God or the Devil, has deucedly damaged the credibility of this codependent Duo.

In any case, to avoid the Devil being recruited by some greedy executive search firm, the Vatican recently expanded his job description so he could exert his vile wiles in more subtle ways. For example, Cardinal Medina - an apparently ***happy and powerful*** *prelate* working at the Vatican - mentioned several devilish deceits that one should always be ready to resist. These include the Devil encouraging men to seek *happiness through power*, or through money or carnal lust - or suggesting that they don't need God or grace or salvation (or the Vatican) - or inducing a vehement aversion to God, the Blessed Virgin, the saints, the cross or sacred things.

As a final example of devilish temptations to avoid, Medina pointed out that "When a child answers the phone and the mother is busy and tells the child to say she is not at home, this is his (the devil's) work". Here Medina obviously meant the lie, not the housework, since any devil willing to do housework would truly be an angel.

Religious fundamentalists recoil from the slippery slope leading downward toward scientific reality

Many pesticides have come and gone in recent years. It has been good business for pesticide manufacturers to exaggerate the safety of their current products until newer, more profitable pesticides became available. Older pesticides with expiring patents were then belatedly recognized as *unduly dangerous* and made illegal for use in the United States. This neatly prevented American farmers from using suddenly inexpensive - because now generic - pesticides that could still be sold to poor farmers overseas for use on the foods we import.

Similarly, gods come and gods go. And it has been good business for priests to routinely misrepresent the god(s) that they manufacture as "hazardous to your life" or "hazardous to your eternal soul" - until more profitable (powerful) gods take over. Of course, unlike patent-expired pesticides, outmoded gods can at least be discarded without endangering the environment. Nevertheless, it is best to discard your outdated gods unobtrusively, for true believers die out slowly and some take their outmoded hypotheses seriously enough to kill for them.

At the moment, the historical credibility of our three great monotheistic religions rests upon the Hebrew Bible. Unhappily, the high price of their codependency, namely, widespread religious warfare between - and unending fraticidal discord within - each of these "fractious by design" belief systems, has not diminished over the centuries.

However, something of fundamental importance finally has changed. And as a result, monotheistic true believers will never again muster sufficient power and resources to conquer, convert and destroy their challengers. So what is the nature of this momentous change, and why don't we see it?

For one need not channel surf very long to locate someone raving about eternal torment or selling salvation. And many politicians continue place *God, The Hypothesis* ahead of duty to family or country, and they take their oath of office *"So help me God!"* with one hand on the Bible. The Boy Scouts still won't admit atheistic boys. The International Red Cross still won't allow Israel to join because those Jews don't want a red (or other color) cross on their affiliated vehicles.

But while Israelis cannot enter under their *Star of David*, Muslim members are welcomed into the Geneva-based Red Cross under the *Star and Crescent* of Mohammed. Of course, Christianity never did depend upon the Koran - and the Old Testament mentions neither Muslims or Christians. Hence only the Jews still threaten Christianity's complex contention that ***Good Christians are now GOD'S ONLY CHOSEN!***

Thus on the one hand, some Christian leaders are eager to recognize a Jew as the AntiChrist. But on the other hand, the Vatican has taken almost 50 years to even partly recognize Israel. For a modern Jewish State with its Capitol in the Holy City of Jerusalem goes against every Christian teaching about *Christians taking the mantle of God's Chosen People from the Jews.*

More specifically, many Christians expect the Messiah to appear in Jerusalem when he returns - and ***soon!*** So for some, the current Jewish State should not have happened. *But no matter whom they may currently assail, despise, impeach, depose or kill, religious conservatives are actually fighting evidence-based reality.* ***And that battle they will never win.***

After all, even the formerly totalitarian Catholic Church - which for over fifteen hundred years has infallibly declared Earth the immovable center of the Cosmos with Heaven up and Hell down while sun and stars traveled across the heavenly firmament each day - has recently had to admit *that Earth is not flat - that Earth rotates as it circles the sun - that the sun is an ordinary star - that both sun and Earth are billions of years old - that there is no absolute up for Heaven or down for Hell* ***- that the evolution of all life forms, including Catholics, is a fact! -*** *that God's role in Creation was not as described in Genesis -* ***and that the Bible*** *"cannot be taken literally" because it* ***is frequently wrong***.

Can wishing it were true restore Biblical infallibility?

When Pope John Paul II decided that Catholic religious belief could no longer contradict evidence-based reality, he knowingly placed the Church on *a slippery slope from which there could be no return to the doctrine of Biblical infallibility.* With that brave act, the pope once again displayed his total faith in the reality of God, in the existence of an immortal soul and in the important role of the Holy Catholic Church. Indeed, other prelates have publicly marveled that "This pope has the faith of a child".

Pope John Paul II felt empowered to deal with this festering doctrinal issue because of his unshakeable conviction that Christian truth - properly understood - could never conflict with evidence-based knowledge about God's own universe (although it surely does). In marked contrast, fundamentalist Jews, Christians and Muslims (those with less faith than the pope) - continue to attack or ignore all scientific findings that show their religion to be based upon outmoded information which always was untrue.

Of course, religious fundamentalists have learned through bitter experience to stay away from science. For it takes a lifetime of hard work to acquire a solid understanding of science. And though relatively few creationists have shown enough proficiency and persistence to acquire a good scientific education, the outcome of that effort has too often been an abandonment of religious fundamentalism.

Yet at some level, even the most hard-nosed Biblical literalists must suspect that the truly important ideas, actors and events recorded in the Old Testament - those upon which monotheism finally rests - are merely deeply engrained superstitions, primitive myths or priestly fabrications with no scientific merit or historical credibility.

Apparently, Pope John Paul II now accepts the basic idea that humans, chimpanzees and gorillas all hail back to the same ape-like ancestors about 7 million years ago - although he has carefully stipulated that only humans can have eternal souls installed by God. But the pope's caveat about the impossibility of souls entering other life forms is an obvious theological non-starter. For in the first place, a soul occupies no space and therefore can fit anywhere God might choose to stuff it - and in the second place, no human can logically postulate any limits on God's infinite creative powers.

Nonetheless, the claim that "ONLY HUMANS CAN HAVE SOULS!" clearly identifies a matter of overwhelming importance to the Vatican. Indeed, this pope's position on evolution in no way downgrades the Church's long term financial prospects as long as unborn souls (potential Catholics) remain within its purview. For not only did the Church reinvent and revitalize eternal souls - its current and future revenues depend almost entirely upon retaining a dominant position in that market.

Therefore, the Vatican will continue to insist - without evidence - that it has God's authorization to set policy on birth control, pregnancy and abortion for everyone on Earth. And despite its acceptance of human evolution, the Church plans to retain its hypothetical God-given exclusive right to ear-mark invisible souls for alleged harp lessons or eternal torment.

But if man truly descends from ape-like ancestors (as the fossil record amply reveals), then:

a) *There cannot have been an original week during which God created the universe and all it contains. Which means*
b) ***there was no Original State of Grace in the Garden of Eden***
c) *Nor any Garden of Eden.*
d) *But had the Garden of Eden existed several million years ago, then Adam and Eve surely would have been very ape-like.*
e) *So modern man clearly is not created in His Ape-Like Image.*
f) *Of course,* ***without an Original State of Grace, the concept of Original Sin loses all meaning,***
g) **hence Jesus cannot have died for your sins** - and then the dominos really start to fall, as detailed by William Jennings Bryan (for entire quote, see page 33):

"...this principle of evolution disputes the miracles, there is no place for the miracles in this train of evolution, and the Old Testament and the New are filled with miracles. If this doctrine is true this logic eliminates every mystery in the Old Testament and the New and eliminates everything supernatural, and that means they eliminate the virgin birth - that means they eliminate the resurrection of the body - that means they eliminate the doctrine of atonement and that they believe man has been rising all the time, that man never fell, that when the Savior came there was not any reason for His coming; there was no reason why He should not go as soon as he could, that He was born of Joseph or some other correspondent and that He lies in His grave..."

h) at which point, ***Christianity has nothing left to offer.***

Chapter Nine

****

"The Washington Post has reported that the Kansas Board of Education may adopt a statewide curriculum that would wipe out any mention of evolution or the origins of the universe and may substitute references to creationism. It further indicates that similar efforts are underway in many other states."

(An American Astronomy Society news release)

****

"Kansas set to banish Darwin from the curriculum. EVOLUTION may be *(indeed, it was)* wiped off the school curriculum in Kansas today in the latest attempt by American "creationists" to promote science based on a literal reading of the Bible. After weeks of skirmishing between advocates of secular and faith-based science, the Kansas school board will vote on whether Charles Darwin should be ejected from the science syllabus in high schools.

"Religious conservatives, who make up half the ten-member school board, argue that Darwin's theory is unproven. They are pushing for a new statewide curriculum that would exclude almost all mention of the theory of evolution, natural selection and the origins of the universe.

"Science teachers would not be prevented from outlining the basics of evolution, but under the new rules students would not expect to be tested on the theory. The move has been condemned by the heads of all six Kansas state universities, who wrote to the school board warning that the exclusion of evolution from the curriculum "will set Kansas back a century".

"Kansas is one of a growing group of states, including Arizona, Alabama, Georgia, Illinois, New Mexico, Texas and Nebraska, where attempts are under way to scale back the teaching of evolution, but the Kansas revision is by far the most radical. "This is the most explicit censorship of evolution I have ever seen," Molleen Matsumura, a spokesman for the National Centre for Science Education, said. "I think other states are looking at what is going on in Kansas."

"Creationists argue that the Book of Genesis should be understood literally and while Supreme Court rulings prevent local authorities from requiring state schools to teach creationism, the anti-evolution movement is gaining ground across America. Polls show that 44 per cent of Americans believe that "God created man pretty much in his present form at one time within the past 10,000 years," according to The Washington Post. Barely one in ten holds the strict viewpoint that life evolved without divine forces.

"In 1998 the National Academy of Sciences fought back with a statement asserting that if American school children are to have any understanding of biology then evolutionary theory is essential. "There is no debate within the scientific community over whether evolution has occurred, and there is no evidence that evolution has not occurred," the academy declared.

"There is plenty of debate, however, within the religious community, with creationists arguing that Darwin's theory not only contradicts the Bible, but is also founded on inadequate and speculative science."

"On the growing battle between creationists and religious conservatives"

London Times by *Ben Macintyre* (*August 11, **1999***)

****

Facts speak for themselves

As the only scientific (evidence-based) explanation for life, modern Evolution Theory soon convinces anyone who is willing and able to understand the evidence. In contrast, our many conflicting religions must constantly promote their controversial convictions merely to retain market share.

It makes no difference if monotheism's "true believers" are just a small minority in their own religious group, as long as they can rouse the rabble to a warlike state whenever the occasion demands. Indeed, that is how wealthy religious leaders have dominated and impoverished ignorant populations for millenia.

As is true with any power grab, religious fundamentalists use fair means or foul to gain and retain authority. So when out of power, they demand freedom of speech, freedom of the press, freedom of assembly, fair elections and equal justice for all - yet they may simultaneously target innocent civilians for terrorist attacks in order to undermine the secular government.

Then, once in office, religious extremists take over the news media for propaganda purposes and oppress the opposition with police brutality, torture and summary executions. Fundamentalist states also support international terrorism wherever that might advance their cause. For when God gives the orders, zealots cannot refuse - nor do kindness, compassion, compromise, common decency or human laws have further relevance.

But the great strength and durability of ***militant*** *fundamentalist* ***monotheism*** lies in its appeal to the lowest common denominator - the ignorant, the hopeless, the angry, the greedy and the vile - with similar religious leaders always lurking to motivate, justify and forgive the most terrible evils perpetrated in the name of Lord.

As one might expect, that sort of leadership power is especially hard to give up, since most of those who incite, perpetrate and profit from high religious crimes and misdemeanors would otherwise be ordinary social misfits with no useful skills, educational attainments or opportunities for advancement.

Monotheism threatens world peace

Much of human history is an agonizing story of poverty, famine, warfare and pestilence. When compared to the nasty, brutish and short lives enjoyed by most of their ancestors, today's Americans appear uniquely blessed with health, wealth, knowledge and peace.

But while our science-based technology-driven civilization has no known limits on its upside potential, human society also faces enormous problems. Indeed, civilization will live or die according to how well it develops that potential and mitigates those serious problems. Unfortunately, religious fundamentalists view these matters quite differently.

So rather than cooperate to promote public education and encourage scientific solutions for overpopulation, AIDS and environmental problems, creationists compete to discredit or suppress the evidence-based scientific insights upon which humanity's future depends - merely because a widespread public understanding of those well-founded insights would finally discredit religious fundamentalism and make it unprofitable to market Bronze Age myths.

Indeed, creationists operate in such a counter-factual state of denial that their major stated goal is to ***reestablish God's Kingdom on Earth!*** Yet if we look back to the *golden days of religious rule* - more commonly known as ***The Dark Ages*** - it becomes clear that God's Kingdom on Earth simply means fundamentalist leaders in positions of great wealth and power while the passive and ignorant populace returns to poverty, famine, pestilence, religious warfare and despair.

The inability of religious fundamentalists to tolerate our rapidly changing scientific and social reality is now a major deterrent to world-wide peace and prosperity.

Disastrous examples of self-serving, anti-democratic, creationist efforts to rule include Afghanistan, Bosnia, Nigeria, Sudan and certain states of the former Soviet Union. *Furthermore, violence and oppression follow wherever religious sects seek power in order to impose their counter-factual hypotheses.* And that is why *exclusionary, monotheism-supported social disorders* continue to threaten or devastate

Algeria, Egypt, Turkey, Indonesia, Israel, Northern Ireland, Pakistan, Kashmir and the Philippines.

In addition, a brief look into any history book - or merely reading the myth-laden pages of the Old Testament - will document that religion-based or religion-enhanced ***warfare, persecution, ignorance and suffering have flourished ever since monotheism was invented.***

Even in these United States, anti-democratic Christian fundamentalist groups affiliated with the Republican Party made major pre-millenial efforts to incite wide-spread governmental disruptions (hoping thereby to encourage a Y2K return of the Messiah?!) when, among other things, they impeached our elected President for denying an extramarital affair. Not surprisingly, these efforts turned farcical as they sought a presumably sinless Republican leader to cast the first stone.

On Biblical Truths and Other Oxymorons

It is unwise to allow a camel's nose inside of your tent as the rest of that smelly beast is sure to follow. *It is a mistake to feed wild bears* as they may consider you part of their picnic. ***And it is a grievous error to politely tolerate the scientifically disproven ideas of creationists*** just because those ideas closely resemble reassuring religious myths inherited by a great many rational people. ***For religious fundamentalists intend to control everyone and everything***.

So while they may sound almost reasonable when lobbying for tax-supported vouchers so poor inner-city kids can attend the public, private or parochial school of their choice - they seem quite unreasonable when demanding organized prayers and the posting of God's Commandments in our public schools - and they become obvious bullies when trying to stifle the best teaching efforts of our public-school science teachers.

And since creationists regularly declare that their ultimate goal is to ***rule*** our society according to strict biblical precepts, ***there can be no question that they despise our basic American values. Or that they hope to eliminate our Constitutional***

rights, get rid of our independent judiciary, and turn our nation into a religion-controlled totalitarian society.

Must religious fundamentalist rule always be totalitarian?

Many fundamentalist monotheists actually demand that ***you*** live by the particular divine rules ***they*** have derived through their idiosyncratic readings of the Bible. Of course, they might as well insist upon ruling the world by reading tea leaves, since creationists cannot agree among themselves on what the Bible truly says or means - or even on which English version is the infallible translation of God's Own Hebrew.

For example, into the late 1990's, many fundamentalists contended that the ***King James version*** was a divinely inspired translation of the Hebrew Holy Book. But then some creationists stumbled onto the well-known fact that King James was gay - so now the search is on for a suitably sinless and therefore infallible replacement.

However, until that allegedly divine source of religious insight is identified and confirmed, a reasonable person might expect creationists to stop making authoritative pronouncements based upon their discredited King James version. An objective person might even anticipate that creationists would withdraw all previous dogmatic declarations based upon such a sinful, hence fallible, source.

Yet reasonable people forget what religious fundamentalism is about - namely, ***winning and retaining power***. So despite our Constitutional separation of Church and State - and the newfound fallibility of their Bible - ***Christian extremists still claim, but cannot document, a God-given right to decide every educational and moral issue in His name***, *as well as a God-given right to intrude upon the privacy of the great majority who believe differently*.

Some examples: Contraceptives only became legal nationwide over strong Christian objections. Christian fundamentalists routinely harass women seeking medical care. Christian terrorists regularly threaten and sometimes kill defenseless clinic personnel for providing legal reproductive services.

Furthermore, various sexual practices between consenting heterosexual or homosexual adults remain illegal in many states at fundamentalist insistence. Yet these sexual practices can only be documented and persecuted by a *peeping cop,* or through *unreasonable search* (Fourth Amendment) or by *compelling self-incrimination* (Fifth Amendment).

To note that Christian and other religious fundamentalists are unwilling or unable to accept modern reality is to belabor the obvious. Indeed, they only engage in discussions where they can control the agenda to attack outsiders or promote the unscientific Bronze Age beliefs of their particular faction.

But politics is the art of the possible. And true believers cannot compromise. So the overt intrusion of religion into politics is an ongoing threat to every form of representative government.

Christian fundamentalists fear unfettered minds

A vocal minority of citizens in this great land are Christian fundamentalists or true believers. These creationists belong to many major and minor religious factions. Each faction quite sensibly views all the others as mistaken and evil. Nevertheless, their leaders cooperate to sustain the bitter antiscientific tone that so often permeates public discussions of evolution theory, genetics, molecular biology, radiometric geology or astronomy.

The genuine frustration and fear that fundamentalists feel when confronted by modern science ought not surprise anyone, since the evidence-based discoveries of countless scientists long ago discredited every Bible tale that once legitimated Jewish, Christian and Muslim fundamentalism. Nor can one logically expect true believers to approach scientific discoveries with an open mind, for if a creationist allows confirmed evidence to uproot even one of his/her irrational beliefs, prejudices and expectations, the rest are likely to topple as well - resulting in one less true believer.

In fact, when a person's life has been permeated by religious thought and prayer from earliest childhood, it is horrifying and disorienting to have that irrational foundation repeatedly challenged. Even the way religious people explain things to themselves - and the manner in which they deal with success or adversity - must be applied rigorously or it soon stops working. Consequently, they cannot accept what scientists have learned without rendering meaningless their most treasured traditions, greetings, dietary routines and daily rituals.

Thus the modern scientific understanding of life and the universe has such disastrous implications for Bible-based religious faith that fundamentalists rightly fear their own unfettered thoughts. Furthermore, if religious fundamentalists were simply to abandon every religious story or claim as it proved logically untenable, they would already be worshipping that markedly shrunken idol known as ***Our God of the Gaps.***

For most scientists nowadays recognize only three significant gaps in the scientific understanding of life and the universe, namely **the basis for the big bang, the origin of life, and the nature of human consciousness**.*

*See also "God's Three Places" on page 119.

But even these three gaps may soon be seen as unimportant. For example, many biologists consider consciousness an unavoidable aspect of intelligence among higher animals. And many computer scientists see consciousness as an inevitable side-effect of information processing complexity. On the other hand, some philosophers still prefer to view consciousness as a manifestation of the human soul.

Yet no matter how anyone may interpret the remaining gaps in our evidence-based explanations, those gaps are shrinking at a remarkable rate as science advances. And already they cannot support any god-like figure except at great mental cost to the true believer - as anyone who tries to engage a religious fundamentalist in meaningful debate soon discovers - for absolute believers are absolutely unable to hear divergent views.

Of course, the fact that devoutly religious persons tend to be narrow-minded is hardly newsworthy. After all, every one of our conflicting creationist religions proudly claims exclusive possession of God's own truth - which logically assures each true believer that his/her own faction's unique religious opinions are the only ones worth studying. Yet if *any* faction of *any* faith had *any* evidence that it represented *any* God, the other feuding fundamentalist factions would surely join up in a heartbeat.

The Tower of Babel story is an interesting Old Testament adaptation of an ancient Chaldaean legend about the birth of many gods and the destruction of an uniquely tall temple mound in early multilingual Babylon. Thus while the storyteller allegedly overhears God plotting with unindicted co-conspirators to promote multilingualism, this tale actually seems to suggest that God underwrote innumerable start-up religions just so men would never be able to cooperate and reach their true god-like potential.

In that case, monotheistic religions may already have been recognized by the ancients as a poisoned chalice - lovingly designed, created and passed along by individuals who consistently placed their own personal welfare above truth or the public good.

Their unconvincing message leads Creationists to violence

Great armies, criminally inclined buccaneers and free-lance vigilantes have always been Christianity's most effective purveyors, especially when introducing the Prince of Peace to new territories with no apparent interest in Him. The currently subsiding millenialist religious rhetoric is therefore true to actual historical experience and should be taken seriously when it suggests that *many true believers are eager to kill all heretics, blasphemers, witches, sodomists, Jews, blacks and other detested minorities as soon as their own Christian fundamentalist faction can gain control of our sinful society* - which confirms the commonly held notion that ***those who***

worship a hateful and mean-spirited God tend to become hateful and mean-spirited themselves.

But on the positive side, the current American search for "something to believe in" by those who have lost faith in an obviously fraudulent and intolerant deity, may actually encourage compassionate changes that are in our society's best interests. For most people feel better about themselves when helping others in distress to deal more effectively with their problems. Note the marked contrast between "helping one another" and the mean-spirited, judgemental, exclusionary Judeo-Christian-Muslim ***fundamentalist*** model that still causes such world-wide distress.

Christianity's biggest doctrinal difficulty was deferred through Faith

The early theological problems facing Christianity were largely unavoidable. After all, Christianity began as a highly modified offshoot of Hebrew monotheism - which in turn succeeded by aggressively altering or deleting ineffective or irrelevant parts of the older Egyptian model. But before long, Christianity's inherent dependence upon Old Testament justification became its most critical - though usually unspoken - theological weakness.

Thus Christian doctrine only makes sense if viewed as a repeatedly amended manifestation of the relentless Christian effort to replace the Hebrews as God's Chosen People. In Christianity's early centuries, this *replacement problem* stimulated many complex doctrinal reformulations. Eventually, however, some unsung marketing genius finally recognized that Christianity's incomprehensible dogma - the inadvertent outcome of all those ad hoc alterations - had actually become its greatest strength.

And from that fabulous but unrecorded "eureka moment" forward, the Church no longer sought to revise or explain its doctrine more logically. In fact, the Catholic Church thereafter discouraged lay Bible studies - insisting that the mysteries of God in Christ could only be reconciled through faith.

We have previously noted that a Good Christian's unbelievable ability to believe the irrational and incomprehensible was early declared a personal gift from God - a wonderful confirmation of His existence, as well as a strong indication of the recipient's worthiness for eternal bliss.

And ever since that stunningly selfish, so naturally self-perpetuating system based upon positive religious feedback was first set in motion, true believers have felt impelled to ensure their own salvation through repeated public professions of personal faith to any innocent bystanders who might perhaps agree or merely were too polite, too intimidated or too busy to contest the relevant issues.

Recall that Christianity's wealth and status depends entirely upon convincing the ignorant masses that their postulated Christian hereafter is far more important than the obvious here-and-now.* Hence these repeated declarations of faith not only helped wavering Christians to sustain their own beliefs - they also reminded *all* listeners ("Are you listening, God?") that such illogical unquestioning souls surely deserved salvation.

*See A World Lit Only By Fire by *William Manchester*.

Even today, many Americans continue to mistake a public profession of Christian faith for evidence of great personal goodness - *though none can explain what Christian theology might actually mean.* But that really doesn't matter! For as Moses demonstrated several millenia ago, monotheistic religious belief depends upon harsh and relentless reinforcement in order to spread and persist.

The Muslim faith is similarly sustained through multiple daily calls to mandatory public prayers performed in a submissive, humiliating posture. But the raw intuitive brilliance of the first Christian priest to recognize and ritualize this self-substantiating circle of repetitive religious declaration and revival has never been equaled. You just gotta believe!

And with that one brilliant stroke, Christianity redefined itself as yet another of God's infinite - and therefore incomprehensible - mysteries. That being the case, nothing could possibly be wrong with the Church's implausible dogma - which in turn implied something was terribly wrong with anyone who didn't believe it.

Thus by making the theological burden abstract - and by transferring the responsibility for providing further answers to God, whom none dared question - Christianity suddenly became the simple, miraculous belief system that credulous multitudes could accept without questioning.

Chapter Ten

****

"... I have always strenuously supported the right of every man to his own opinion, however different that opinion might be to mine. He who denies to another this right, makes a slave of himself to his present opinion, because he precludes himself the right of changing it.

"The most formidable weapon against errors of every kind is reason. I have never used any other, and I trust I never shall...

"Your affectionate friend and fellow-citizen,"

Thomas Paine (1794)

(from the introduction to The Age of Reason
"Being an investigation of true and fabulous Theology")

****

Christianity's dependence upon the Hebrew Bible led to persistent persecution of the Jews

Whenever possible and practical, the Church reemphasized the importance of faith to all doubters by threatening and isolating (in the case of Galileo) or simply killing them (e.g., in the cleansing flames of the Holy Inquisition). For the Christian hierarchy soon recognized - as wise ayatollahs still do - that a religious dictatorship cannot long endure if the faithful feel free to disagree, dispute, disengage and disappear in pursuit of more palatable philosophies.

The fact that no fundamentalist religion has convincing evidence to support its pomp, circumstance and outrageous demands, is regularly reconfirmed by the perpetual disputes between all monotheistic religious factions, sects and cults. And clearly, open disbelief could never be tolerated by those peddling implausible nostrums like the following:

Jesus Christ was God's own Son, sent here to atone for mankind's sins. Mankind's sins vaguely refer back to long, long, long ago (talk about holding a grudge!) when Eve may have tempted Adam to sin - either by offering a bite of her apple (Egads!) or by helping Adam to gain sexual knowledge (Bad Girl!). Or perhaps the real sexual sin was something else entirely that the all-male Church leadership has yet to confess or confront.

Keep in mind that the One True God who designed Adam and Eve, snakes and apples, could as easily have created a pair of ape-brained apple-averse eunuchs and thereby saved the world endless grief. Anyhow, according to some strange (un)Godly logic, God finally figured out that He could only save suffering humanity from His Own Great Wrath by getting some poor married virgin pregnant in order to later kill their Perfect Son as a token of His Great Grace - or something like that.

Well, you have to credit the Christian God for managing to suppress that persistent pique over a stolen apple or teen age sexual explorations or priestly perversions for several thousand years. But eventually He just had to kill someone. Of course, this entire horrific and implausible Christian libel - that any rational being would sacrifice His own son to Himself in order to absolve sinful humanity from His own great wrath - was the inadvertent culmination of an incremental series of increasingly illogical efforts to legitimate the struggling Christian Church.

But given the Church's obvious conflict of interest in devising such a tormented tale, the entire story was hastily repackaged as a mystery - then publicized with profound expressions of gratitude for God's Great Grace in thereby ensuring a permanently profitable priesthood. For then it came to pass that personal salvation could only be obtained through the Church - which proudly declared itself God's newly appointed Representative on Earth as soon as word got out that God had betrayed and murdered His own Son to partially compensate Himself for Mankind's Original Sin.

Yet strangely enough, we see no evidence that the shabby way God treated His Own Son *who-was-perfect-in-every-way* in any way worried the *way-less-than-perfect* Christian leaders who have been fighting over Jesus's lamentably vacated franchise as God's Earthly Rep ever since. Indeed, no sooner were their initial sectarian battles behind them, than those early Primates celebrated their victories by completely inverting the awesome responsibilities they had allegedly inherited from Jesus.

No longer would they chastise the rich and run money lenders out of the Temple! For now they were proud to bear the title of ***God's Officially Designated Collector of Taxes and Bribes***. Of course, those bribes were routinely misidentified - some arrived as "Donations to God in Honor of His First Five Thousand Years as ***Overseer of Preferences and Punishments"*** - others were simply titled "Alfredo wishes God a nice day", or whatever.

Some might wonder why Christian theorists chose to reinvent, rewrite and reverse God's ***P*** *and* ***P*** *rules - given that Jesus supposedly had clarified and codified them - especially as Preferences and Punishments allegedly justified God's Vast Eternal Plan from the start.* Nonetheless, with God's example before them, it soon became necessary to bribe all Church officials to perform their divinely defined duties. Which brings up endless questions such as:

Do those who donate to religions really expect their donations to alter God's fair and impartial verdict? Or do priests merely contend that the money they accept for hypothetical services not rendered will buy P and P favors from God when they fully intend to keep it for themselves?

Christians have always tried to displace God's Chosen

We have seen that the implausible twists and turns of Christian doctrine can only be understood as an inadvertent outcome of repeated Christian efforts to indirectly replace the

Jews as God's Chosen People. For Christianity could not directly deny and usurp that Hebrew status without thereby undermining the credibility of the Old Testament upon which Christian legitimacy finally depended. ***The unavoidable illogicalities of that theological dilemma underlie almost two thousand years of Church-sponsored anti-semitism.***

The rationale for Christian anti-semitism remains inescapable:

"We Christians have inherited, and now depend upon, the Hebrew Holy Book. Because we are strong - and 'might makes right' in theology as in all else - all Jews must convert to Christianity and/or die" (as did most native Americans and countless others). But the Jews never relinquished their patrimony as God's Chosen People, nor were they willing to convert or die quietly so that Christianity might make more sense.

Therefore, every generation of Christians has included many true believers who - as a favor to the Church and to ensure their own salvation - have tried to resolve Christianity's Theological Dilemma by killing Jews whenever possible. Indeed, an organized massacre of Jews became such an integral aspect of Russian Christian life that it was commonly referred to as a *pogrom.* This naturally brings up Hitler's Final Solution to the *Jewish Problem.*

Though the exact nature of that problem was never made clear in public, the Final Solution soon became known as the *Holocaust* - a burnt offering to God. Of course, Hitler's decision to kill all Jews as a favor to God and the Church was a natural outgrowth of his good Catholic upbringing - during which he first became a communicant and then an altar boy before finally being confirmed as "a soldier of Christ".

Throughout that standard religious preparation (for what?), Hitler was routinely taught to despise *the perfidious Jews.* In his later years, Hitler also became a great admirer of Martin Luther (the founder of Lutheranism, Germany's other major religion) who equally despised all Jews. Eventually Hitler wrote in *Mein Kampf* "I am convinced that I am acting as the agent of our Creator. By fighting off the Jews, I am doing the Lord's work."

In 1941 Hitler declared to Gerhart Engel, “I am now as before a Catholic and will always remain so.”*

*Quoted by John Patrick Michael Murphy in “Free Inquiry” (spring 1999).

Hitler’s proposed destruction of the Jews - as well as his prohibition of abortion and his dutiful collection of taxes for the Church even during wartime - obviously served critically important Catholic interests. As a result, the Catholic Church quietly supported his goal of a world-wide *Christian* German Empire.*

*See also Hitler’s Pope by *John Cornwell.*

Furthermore, Hitler’s All-Christian Nazis and their many non-German Christian colleagues - for example, the Catholic Croats of Yugoslavia - felt duty-bound to kill as many Orthodox Christian Slavs (e.g. Serbs) as possible; the unforgiveable sin of those poor misguided Slavs being to follow the Orthodox Christian Patriarch who was the Catholic Pope’s only equally qualified Christian competitor for the post of God’s Earthly Rep.

Christian belief grows from the barrel of a gun

Interestingly enough, despite their long fraticidal relationship, the patriarch’s and pope’s negotiators recently nearly succeeded in recombining the Orthodox and Catholic Churches. But while they easily resolved all religious and business issues - rituals, parking spaces, finances, marketing and so forth - the negotiations broke down when neither pope nor patriarch would take orders from the other guy.

Some might claim that both prelates thereby revealed an unseemly lack of Christian humility. Others merely deplore the incessant infighting over who really represents the Prince of Peace. For this self-serving behavior by top Christian leaders is what condemns poor uneducated Serbs and Croats - good

Christian soldiers all - to continue killing one another whenever they can spare time from murdering Muslims or Jews.

Note that the so-called Jewish Problem was actually a Christian *theological* (i.e., purely hypothetical) problem. However, Church apologists have always preferred to divert attention from Christianity's obvious theological deficiencies by labeling the Jews *just another unlovable, easily scapegoated minority* - comparable, for example, to the unpopular Gypsies who also took a hit from the Nazis.

That the *Final Solution* was military is hardly surprising either, for Christianity has always spread primarily through governmental fiat and military conquest rather than via logically convincing discourse between calmly objective adults. In other words, Christian faith grows from the barrel of a gun. It is spread by force.

And every Christian military victory reconfirms the fundamental efficiencies of ***monotheism**, namely*

1) *Never having to share the wealth with representatives of other gods,*
2) *nor to tolerate a second opinion from representatives of other gods,*
3) *nor (before Pope John Paul II) to apologize for all the evil done*
4) *"In The Service of The Lord".*

In a final paradox, Christianity presents itself as the religion of hope and resignation. And among the most devout Christians are those whom Christianity has made most miserable by religion-supported warfare - by religion-supported dictatorships - by religion-supported intrusions into their personal affairs - by religion-supported confiscations of private property - and by the religious education that so routinely instills incapacitating guilt that only the Church can partially absolve, for a price.

Summary: Christianity arises from the barrel of a gun *(might makes right).* **It depends upon *faith to confirm theology*** (a dizzying example of ***circular reasoning***) - and its greatest appeal is to its countless victims - the totally defeated - those whose former gods proved powerless to protect them - the impoverished - the overpopulated - the enslaved - and their uneducated (at least in science) descendents.

But all is not lost. For to repeat Thomas Paine's astute observation, *"Reason and Ignorance, the opposites of each other, influence the great bulk of mankind. If either of these can be rendered sufficiently extensive in a country, the machinery of government goes easily on. Reason obeys itself; and Ignorance submits to whatever is dictated to it".*

In other words, a society will naturally outgrow the survival benefits of blind patriotism and religious faith as its members become educated in science and accustomed to reason.

Of course, the inevitable switch from a myth-based to a reason-based society is bound to elicit tremendous opposition from the ignorant and those who profit by keeping them so. And that is where we are today.

Chapter Eleven

The search for meaning in an age of reality

****

A final repeat of the Christian Fundamentalist position

"Your honor asked me whether evolution has anything to do with the principle of the virgin birth. Yes, because this principle of evolution disputes the miracles, there is no place for the miracles in this train of evolution, and the Old Testament and the New are filled with miracles. If this doctrine is true this logic eliminates every mystery in the Old Testament and the New and eliminates everything supernatural, and that means they eliminate the virgin birth - that means they eliminate the resurrection of the body - that means they eliminate the doctrine of atonement and that they believe man has been rising all the time, that man never fell, that when the Savior came there was not any reason for His coming; there was no reason why He should not go as soon as he could, that He was born of Joseph or some other correspondent and that He lies in His grave; and when the Christians of this state have tied their hands and said, 'We will not take advantage of our power to teach religion by teachers paid by us,' these people come from the outside of the state and force upon the people of this state and upon the children of the taxpayers of this state a doctrine that refutes not only their belief in God but their belief in a Savior and belief in heaven and takes from them every moral standard that the Bible gives us.

"...I have been so well satisfied with the Christian religion that I have spent no time trying to find arguments against it. I have all the information I want to live by and die by."

William Jennings Bryan (at the Scopes trial, 1925)

****

in order to compare it with the scientific or secular position

"The state of Tennessee, under an honest and fair interpretation of the constitution, has no more right to teach the Bible as the divine book than that the Koran is one, or the book of Mormons or the book of Confucius or the Buddha or the Essays of Emerson or any of the ten thousand books to which human souls have gone for consolation and aid in their troubles.

"I know there are millions of people in the world who derive consolation in their times of trouble and solace in times of distress from the Bible. I would be pretty near the last one in the world to do anything to take it away. I feel just exactly the same toward every religious creed of every human being who lives. If anybody finds anything in this life that brings them consolation and health and happiness I think they ought to have it. I haven't any fault to find with them at all. But the Bible is not one book. The Bible is made up of sixty-six books written over a period of about one thousand years, some of them very early and some of them comparatively late. It is a book primarily of religion and morals. It is not a book of science. Never was and was never meant to be.

"They make it a crime to know more than I know. They publish a law to inhibit learning. This law says that it shall be a criminal offense to teach in the public schools any account of the origin of man that is in conflict with the divine account that is in the Bible. It makes the Bible the yardstick to measure every man's intelligence and to measure every man's learning. Are your mathematics good? Turn to I Elijah ii. Is your philosophy good? See II Samuel iii. Is your chemistry good? See Deuteronomy iii 6, or anything else that tells about brimstone. Every bit of knowledge that the mind has must be submitted to a religious test.

"If today you can take a thing like evolution and make it a crime to teach it in the public schools, tomorrow you can make it a crime to teach it in the private schools. And the next year you can make it a crime to teach it in the church. And the next session you may ban books and the newspapers. Soon you may set Catholic against Protestant and Protestant against Protestant and try to foist your own religion upon the mind of man. If you can do one you can do the other. Ignorance and fanaticism is ever busy and needs feeding. Always it is feeding and gloating for more. After a while, your honor, it is setting man against man and creed against creed until with flying banners and beating drums we are marching backward to the glorious ages of the sixteenth century when bigots lighted fagots to burn the man who dared to bring any intelligence and enlightenment and culture to the human mind."

Clarence Darrow (at the Scopes trial, 1925)

****

It has been 75 years since William Jennings Bryan and Clarence Darrow delivered their historic summations of the creation/evolution controversy. Yet astonishingly, the debate remains stalled exactly where they left it - with the same heat, the same outrage, the same arguments - only the actors have changed. How could this be?

How could the most momentous religious and philosophical arguments about life and the universe remain unaffected while every field of science has been overturned and revitalized? How could an unprecedented century's relentless cascade of new information ***not*** *have influenced the ongoing dispute between creationists and evolutionists - except to harden both sides in their original positions?*

How could the rising tide of scientific information lift every socially and technologically prepared society to unprecedented new heights of wealth and power - how could that flood of new knowledge swamp and sink so many traditional cultures around the globe - and still have not the slightest influence on the way fundamentalist Jews, Christians and Muslims interpret reality?

Even a reasonable person might begin to wonder whether our modern information-driven society has finally run into something basic and unchangeable. Didn't the ancient Greeks argue over these same issues? Do we actually understand the big picture any better than they did? Might we be overly focused upon unimportant details even now? What if the modern scientific view has been oversold, like so much else in our market-driven economy?

But in that case, why would our lives be changing so swiftly? Why would smart, well-educated people still need to learn so many new things in order to get, keep and do a good job? What other explanation might there be for all the running we do nowadays merely to remain in place? How could regular job retraining be nothing but a fad? What about all of our knowledge-based industries? And why would computers be so significant?

More importantly, is the modern information era truly unprecedented? Or are we simply passing through another cyclical event? In other words, do eternal verities really exist? And if so, where shall we find them and what might they be? We will finish our search for truth among evolutionists and theologians by making a few ordinary observations - first about modern science, and then about modern religion.

Modern Science is cumulative and pervasive

One thing cannot be denied: Science has had an enormous world-wide impact. From antibiotics to global positioning satellites to lasers to television, almost every human being - indeed, nearly every life form - has been affected by scientific advances. Nor can anyone deny that ***modern science and technology have converted public education into a more important national resource than coal or oil or iron ore.***

In fact, no nation can compete internationally without a well-educated work force. But while every investment in public education increases each citizen's value to the ruling class, it also produces an increasingly aware "show-me" public that inevitably develops greater personal expectations and less tolerance for traditional social controls and non-elected leaders.

As a result, the rise of evidence-based science has been associated with a decline in dictatorships, which remain unchanged only in the most backward and intolerant, myth-dominated lands.

In the good old days, life was a zero-sum game with wealth, power and reproductive success dependent upon finite resources such as land, cattle, women and metals - coveted assets that could be acquired by force. But modern wealth is mostly information-based, and informational assets have an irresistible tendency to replicate, spread and become freely available.

Furthermore, useful discoveries may arise anywhere and are easily confirmed. But while information is essentially uncontrollable, it has no value for those unable to utilize it. *One cannot simply grab information by the hair and drag it back to the cave kicking and screaming.*

In addition, modern wars are no longer won by the side with the most peasants to spare. Instead, warfare has become highly efficient and technological - hence prohibitively expensive and destructive of both fixed and fluid assets. So while war profiteers will always be with us, war is no longer a practical way for ordinary nations to accumulate wealth.

At the same time, the widespread dissemination of additional information - and the ready availability of inexpensive technology for reorganizing and extracting new value from all information - has opened unprecedented opportunities for the creation of wealth. Directly or indirectly, new or newly useful information enriches almost everyone. And as that rapidly expanding information-base recruits ever more opportunists world-wide, they in turn push information growth along at an ever-accelerating pace.

It is worth repeating that important new information can no longer be suppressed by dictators or ideologues. And given that the scientific information genie cannot be ignored, forgotten or stuffed back into its bottle, the impact of scientific advances is unlike that of any other cultural development. For with every discovery a potential springboard for launching additional investigations, science inevitably drives human history outward in uncharted directions rather than sustaining the status quo or allowing history to enter repetitive cycles.

Of course, with no one even nominally in charge of scientific advances, no one can possibly predict where human civilization may be headed. There are some, especially the young and the bright, who find this unpredictable rush of events tremendously exhilarating and profitable. But many conservatives lose their bearings and become disoriented or distressed by the rapidity of social and technological change - and by the way new discoveries regularly invalidate "established truths" or move them aside as hopelessly outdated.

After all, most of us work within and profit from the status quo. And even those who refuse to compromise or cooperate eventually become codependents of the world that is. So sooner or later, a significant minority of any population feels forced to "take a stand" - to defend their sunk costs - protect their investments. In a word, they become more selfish, comfortable, conservative and fearful of further change - the stance that William Jennings Bryan articulated in the quote at the start of this chapter.

What the fuss is really about

If we reexamine Bryan's and Darrow's classic statements, it becomes apparent that while facts and theories were bandied about or trampled in the dust, ***the real battle between theology and science was, is, and always will be about power.***

Basically this is a control issue. And each faction quite correctly fears what the other side will do if it can. So like every contest between incompatible world views, it eventually boils down to who wins and who loses, rather than whose theological or scientific assertions have greater merit.

Our conflicting religions are many and subjective. And each new religious start-up must differ, adapt and postulate wildly until it either shrivels and dies or finds its own *sweet spot* ***- some new interpretation of the incomprehensible that resonates -*** something sufficient to support yet another fairy castle of fictions and justify one more profitable belief system and power structure.

In contrast, while various scientists have surely exhibited every well-known human weakness as well as many frailties less apparent, ***science remains a single self-correcting evidence-based process with no apparent limits on its explanatory powers or productivity*** - and all the theology in the world will never alter a single, amply confirmed scientific finding.

As a result, each side enters the fray knowing full well that the other side cannot be convinced - that their opponents must be converted or coerced - by reeducating their children if possible - through political clout or legalistic maneuvers if necessary - or (for the religious fundamentalists who are finally losing their battle with reality) via demagoguery and violence.

So the simple fact that this is a fight for power explains why - after 75 years of unequalled scientific progress - the two sides have not budged. It also explains how they can continue to draw opposite conclusions from the one historical truth on which both sides agree, namely that ***Three great monotheistic religions derive their primary historical justification from the Hebrew Bible.*** For this fact alone forces all fundamentalist Jews, Christians and Muslims to insist that

The Hebrew Bible represents God's Inerrant Word!

Chapter Twelve

Does The Hebrew Bible represent God's Inerrant Word?

In order to make that claim, creationists must ignore or deny countless scientific and historical studies that have consistently revealed the Hebrew Bible to be riddled with myths, misconceptions, inconsistencies and outright fabrications. So how on Earth can religious fundamentalists still insist that God and His Bible are inerrant after so many proven errors?

Well, God could still be infallible if those who retold the Bible stories as oral history - or who finally wrote down them down - were either very careless or promoting their own agenda rather than God's. But regardless of whether most Biblical blunders reflect ignorance, innocent mistakes or deliberate falsehoods, every monotheistic religion based upon an untrue Bible must be false. And the same sad conclusion holds if we test the possibility that God was not infallible, or that the Bible was not His Word.

Indeed, the only way the Bible still might deliver God's Inerrant Word truthfully would be if the scientific knowledge that underlies our modern civilization were untrue - but then scientific advances such as atomic power and space travel could not be. Here an obvious question comes to mind: "With so much evidence available, why hasn't some objective panel of judges or a jury of our peers already settled this long festering disagreement?" For surely the matter has never been inconsequential for our society as a whole.

After all, without the strong incitement of our mean-spirited and conflicting monotheistic religions, why would anyone want to shoot abortionists or persecute gays or finance religious terrorists or carry out ethnic cleansing? Hence if it ever comes under Supreme Court jurisdiction - and it probably never will - the most pressing question before that Court surely ought to be: ***Was the universe created in accordance with the Hebrew Bible or not?***

And as anyone who paid the slightest attention to their high school science teacher already knows, every available bit of objective scientific evidence reveals the same thing - that the visible universe has nothing in common with the Creation Story in Genesis that was written for certain Bronze Age shepherds by their priests.

So why can't our Supreme Court simply hear out the evidence and then ***vote on whether our entire information-laden universe is true or just God and His Bible are true?*** *After all, these are the only logical options.*

For if the universe is partly false or God is partly false, then evidence-based self-correcting science is still our only hope for discovering the true nature of reality. And since a mean-spirited God who is neither infinite nor infallible is merely a loose cannon that cannot be secured, we may as well ignore Him and hope that He remains away on more important business.

It stands to reason that neither all-knowing God nor innumerable misguided angelic employees could possibly have made enough errors or off-plan changes to account for all the differences between what is described in the Bible and what we see. Hence the Court's decision ought not tax their combined capabilities to determine such matters beyond any reasonable doubt.

Other possible explanations - all-knowing God didn't know what He was doing - or He didn't care enough to provide a corrected proof of His Bible - don't add anything new. And following a fair trial of this question, anyone wanting to ride creationism or some similar monotheistic religious fantasy to fame, fortune and political power would simply be out of luck.

Yet when their arguments fall short, as inevitably they must, fundamentalists look truly perplexed and ask, "But wouldn't you rather have Almighty God in His Heaven than believe that life ends with death?" As if a) wishes - despite being notoriously ineffective in this world - might somehow create a worthwhile God in the heaven of their dreams; and b) as if the infinite and eternal God described in their Bible wasn't a far worse nightmare than termination-at-death could ever be.

Fundamentalists frequently fend off scientific criticisms of their magical monotheistic beliefs by insisting that ***rationality and religion are incompatible*** - which appears amply evident to most scientists as well. Nonetheless, ***many religious persons who respect science still seek a common ground between our evidence-based scientific understanding of life and the universe, and their own magical Bronze Age beliefs. But the only common grounds imaginable between fact and fantasy must be imaginary as well.***

On the mysteries of religious belief - what might these be?

Every successful monotheistic belief system goes through an initial period of intense theological explanation and experimentation, test-marketing on focus groups, revision and retuning. In their ongoing effort to minimize disruptive illogicalities that might disorient those entering the faith during this early phase of revision and reconstruction, new religions always insist that only a few of their top minds can endure exposure to certain *divine mysteries* from which the ignorant masses must be protected.

"Some day you may appreciate why we demand this and insist upon that, for it would endanger the sanity of an unprepared mind to learn these mysteries! We keep you in the dark for your own good! Just do as you are told! (After all, if you really understood what was going on, we would have to kill you!)."

Except during carefully monitored market tests, early monotheistic systems have usually released as little information as possible about their evolving theological strategies. Indeed, fifteen centuries passed before the Catholic Church finally allowed the Old and New Testaments to be issued in languages other than Latin or Greek. In the meanwhile, only Catholic clergy were educated or allowed to study theology.

Some historians mourn the waste of those fifteen centuries during which the wealthy totalitarian Church oppressed the poor and ignorant masses through its monopoly on human knowledge. Others of the faith prefer to remember Catholicism as an isolated beacon of light during the Dark Ages, while ignoring the Church's role in bringing about and sustaining those Dark Ages.

However, the Church's harsh dominion over European minds eventually cracked when Martin Luther, John Calvin and other priests broke away from the Catholic old-boy network. Those early Protestants were able to to survive their rebellion without being consigned to the flames because they rode a popular wave of disgust with an entire millenium of Catholic taxation, greed and publicly sinful behavior. But it is hard to give up bad habits, so Luther and Calvin both remained ready to torch anyone who disagreed with them.

Nevertheless, the first Protestants of that Religious Reformation mainly wanted to escape from papal rule (which was widely detested) in order to peel away the overlay of Catholic misinformation that obscured the purity of Christ's life and thereby expose the true principles of Christianity.

Fortunately for them, the Religious Reformation began at a time when the printing of books and the translation of the Bible into vernacular languages was becoming possible as well. Yet many Protestant scholars who began to study Holy Scripture at this time eventually were burned at the stake when their ideas and discoveries went beyond what their former comrades were prepared to accept. Soon a number of mutually intolerant Protestant groups set up their own competing theocracies.

In fact, only now - after five centuries of assiduous and often hazardous investigations - has the relentless Protestant quest for the real Christian Truth finally borne fruit. And today we can not only place the historical Jesus in his own time, but even examine samples of his teachings.*

*See The Lost Gospel by *Burton Mack.*

Thus the good news is that the devout and painstaking process of carefully peeling fifteen centuries of Catholic theological amendments away from early Christianity - as one might peel away the layers of an onion - has now been completed.

But the bad news is that after removing all the layers of unjustified theological assumptions and self-serving religious declarations that have so long obscured the postulated divine center of that onion, it turns out that nothing of theological significance remains to be studied.

Thus the bloody process of religious revival begun by early Protestant theologians - who were convinced that a corrupt Church had sinfully obscured the real Jesus and His message - has inadvertently disclosed that the entire theological enterprise known as Christianity contains no heart of sacred truth.*

*See The Bible As It Was by *James L. Kugel.*

In other words, the Divine Central Core of Christian Truth is hollow. Nor can one dig any deeper. It is this realization that now burdens those who teach and study Christian Theology in Divinity Schools. As a result, most divinity students who accept the modern interpretation become non-believers. Yet many of those non-believers eventually go out to serve conservative congregations that would be outraged if they knew their own minister shared that modern, tolerant understanding of Christianity.

At the same time, over on the traditional side, proudly ignorant fundamentalists continue to trumpet their conservative and highly profitable Bible-based fictions as if nothing had changed under God's heavenly firmament - as if a flat or slightly bowed Earth was still divinely fixed at the center of the universe. And those same outmoded, antiscientific beliefs will surely circulate throughout the southern Bible Belt of the United States until sweaty evangelists no longer can profit by selling tall tales to an increasingly educated public.

In the meanwhile, rather than let their own ***religious beliefs*** waver because they ***violate common sense, ordinary experience and plain human decency*** - true believers proudly compete to see who can swallow greater quantities of counter-factual religious "truths" in one gulp - as if personal salvation was the ultimate prize of a "divine beer-drinking contest".

And surely the most God-fearing, judgemental and unforgiving Southern whites will never consider the possibility that their own intense religiosity may in part be an unintended consequence of agreements made by non-believing planters with the preachers they hired to keep their slaves pacified.

For persistent tales suggest that these planters initially joined the rest in loud and enthusiastic worship in order to lend Christian services greater credibility amongst their illiterate slaves - and that their own regular public declarations of profound faith may inadvertently have convinced many community leaders - as well as their descendents and slaves.

The Secrets Are Out

So what were those closely held secrets of Egyptian monotheism that only a select few initiates might be taught after years of study? What hidden messages did Jewish mystics and Christian preachers seek so avidly and in vain over the centuries? What might the most profound mysteries of any religion eventually reveal about life, death and a hypothetical afterlife?

Well, some claim that these mysteries incorporate powerful and dangerous secrets of pagan magic rituals - others insist that they embody the magical spelling of God's name, along with the dreadful powers and punishments associated with its use and abuse - still others declare that these mysteries include directions on how to reach Seventh Heaven where God rules on high, dazzling the saints and angels with His brilliance.

But it seems far more likely that the great mysteries of monotheism really consist of mundane information - ordinary truths that common people would not appreciate - such as the fact that the ignorant really want to believe in something - or that large numbers of uneducated people tend to become quite unmanageable unless they have shared beliefs about proper behavior that are reinforced by marvelous and frightful tales of divine rewards and punishments - *or that the doctrines of wealthy and powerful religious systems must always be adjusted cynically and cautiously with an eye toward long-term profits, rather than openly in the pursuit of divine truth.*

In other words, the real secret seems to be that any religion's major task is to baffle everyone with its bullshit. And that it is critically important to maintain a unbroken succession of top insiders who realize that these religious deceits (otherwise known as mysteries) are the only way they can protect their highly profitable belief system from unintentional doctrinal damage.

Interestingly enough, Pope John Paul II may inadvertently have inflicted just this sort of collateral damage upon the Catholic Church when his "faith of a child" led him to

1) agree that Earth is not flat, not recently created and not the center of the universe
2) admit that Evolution Theory is irrefutable, and
3) apologize for the well known fact that the Church has prospered mightily through many evil acts.

Only time will tell.

Let us now seek today's real Christian Truth

Most Americans claim to be *Christian.* Some of these Christians belong to one or another of our countless conflicting religious factions. Others don't. Apparently, the *Christian* label may mean anything or nothing at all. Thus some Catholics revere the pope while others find him irrelevant. And many who respect the pope's teachings, simply ignore the lessons they find inconvenient.

Some Protestants think the pope is a fine fellow. Others insist that he is the AntiChrist. Certain Episcopal leaders are convinced of Christ's divinity. Others bishops are agnostic. Many Orthodox Christians submit to the Greek or Russian Patriarch, yet maintain divergent religious views.*

*See also "It was wrong to shoot the pope" on page 128.

Devout Lutherans, Mormons and Southern Baptists have contentious fundamentalist and liberal factions that rarely agree on anything. Christian Scientists, Jehovah's Witnesses and Seventh Day Adventists have little in common with each other or with the larger Christian denominations. Non-believers are common, yet most non-believing "Christians" might agree that "there are no atheists in foxholes".

These confusing realities of modern-day Christian thought encourage most of the devout to ignore doctrinal disputes and deal with such incomprehensible questions in a more personal way. Hence some simply delete most members of competing denominations from their own short list of "Good Christians". And even in the home congregation, many may fail to make that final cut.

Here again, we encounter a lot of diversity, a lot of imagination, and absolutely no concern for current research or historical evidence. For example, Michael Farrell, editor of the National Catholic Reporter, sponsored a contest for artists world-wide, offering a $2,000 prize and a place on the millenium issue cover of his newsweekly for

"a bold new image of Jesus"

to mark the 2,000th anniversary of his birth. The contest welcomed all visual media: computer art, stained glass, silk screens, even photographs.

"The only sure bet is that the winning entry won't resemble the traditional images of Jesus evoked by artists of the past...If you are giving us a repeat of any of those images, it is not likely that you are on the winning ticket. There ought to be something new that we have never seen before."

Yet what else would you expect in the land of the free and the home of the brave? After all, marketing was practically invented here - which is not so surprising either, given that most of us are so hurried and harried by a wealth of personal obligations that we routinely worship without understanding theology, vote without listening to the candidates, pay taxes without comprehending tax law, and just pray that our children will stay out of jail.

Summary: It seems unlikely that we will stumble upon fundamental truths by observing the conflicting practices and beliefs of ordinary Americans - and this may well be our salvation. For just as humanity could hardly abide if everyone desired the same spouse - and our economy would wither if everyone insisted on purchasing the same stock - we could surely kiss freedom and progress goodby if everyone adhered to the same theology.

Thus it is ***a blessing*** *that our countless religious and scientific ideas must vie, evolve and die. It is* ***a miracle*** *that information moves so freely and that evidence can be openly contested. And it is* ***a godsend*** *that no one has any final answers*.*

May the search never end!

* See also “Nights Alone” on page 126.

RELIGION (from A to Z)

Abortion and the human soul

When a sperm meets an egg
'tis a wondrous event
that deserves a new soul
which from Heaven is sent

Now a soul's not an object
that adds to your weight
or a thing that could split
into four or just eight

Yet of six billion humans
one billion might be
fertile and female
with eggs to set free

With each woman including
a large egg account
some hundreds of thousands
in total amount

As for billions of men
'tis no problem at all
to make sperm by the trillions
'tween life's spring and life's fall

Since the soul's an idea
that can't even be measured
this surely implies that
it ought to be treasured

At least by the Church
which the soul first invented
so that all of your sins

might be fully repented

by donations ample
and guilt most sincere,
since an absence of soul
meant no afterlife fear

Now each soul is a handle
the Church has on you
to jerk you around
if you fail to come through

Thus God doth create
countless souls for each day
so the sperm with an egg
gets ensouled as they lay

Does each soul have an earmark
or bar-code to tell
if it's destined for Heaven
or going to Hell?

Could it be that there's really
no difference at all
between souls for the rich
and souls for the small?

If the same soul's inserted
to each egg in the line
then "one size fits all"
should work perfectly fine

But why would God bother
if all souls are the same?
Is your soul just a pawn
in God's favorite game?

If a starter soul enters

each egg where it's hit
in position uncertain
did yours land where you sit?

Or could it reside
in appendix or colon
in which case let's pray
you may never have hole in

the structure that holds
all your assets eternal
whose removal might save you
from fires infernal

For Good Christians would curse
your appendix removin'
if the outcome could be
an eternal soul losin'

How could souls be of use
when they've never been seen
and they have no effects
I would bet they are green

But an egg is so tiny
and easily lost,
every soul must be smaller
regardless of cost

Yet a soul can't be protein
or carbo or fat,
since it must resist burning
if it comes down to that

For each soul should feel pain
if in Hell's fires basted
or love music from Spain
so Salvation's not wasted

A soul must be able
to see divine light,
it must also be small
and stay curled out of sight

Perhaps it's transparent
like small jellyfish
or quicker than light
though you look where you wish

So how can you tell
if your soul is all right?
Could it easily flee
in the still of the night?

Indeed are we sure
that the soul can exist?
Or is this a rumor
passed down through the mist

from ages when humans
with stone axes slew,
and had no idea
what science might do?

Still three fourths of all embryos
die before birth
Can their souls be recycled?
Have used souls any worth?

Might identical twins,
even triplets or four
get one soul for the bunch
or would Great God send more

if a fertilized egg
later split into three?

Might the soul undivided
then even break free?

Delete its bar code
without taking a name?
Leave triplets behind
with no soul in the game?

But if souls have no substance
and have never been met
despite endless soul-searching
is it time to forget

All this Stone Ager nonsense
and start talking clearly
of pregnancies wanted
and environments merely?

For by restricting discussions
to things that exist,
we could argue far less
and no longer get pissed

About things that could be
but apparently ain't
like those heavenly hosts
and the soul of a saint

****

Belief is a costly crutch

Some think of their faith
as a framework internal
that helps them resist
countless pressures infernal

Others limit their thoughts
with a shell of belief
since permitting no doubt
brings a form of relief

Some suffer great pain
over choices gone wrong
and seek to remain
under tutelage strong

Religions are primitive
based on emotion
external controls
o'er internal commotion

Some find it too scary
to rule their own lives
so they seek out a leader
and trust he is wise

Religions feed hopes
that need ne'er be fulfilled
they dissipate energy
which could otherwise build

an existence more fruitful
enriched by great love
with each serving their neighbors
not Heaven above

Religions gain power from
Heaven and Hell
devotion, damnation
and soldiers as well

Religions must dominate
those who now live

since the dead are all rotten
with naught left to give.

****

Closure - as in "Slowin' down"

(to an old friend contemplating suicide)

When time was forever
and each day brought surprise
we hurried to meet
all our goals 'neath far skies

Work daytime, work nighttime
it all seemed the same
and no one got tired
till the end of the game

Now I'm walking so slowly
she can hardly keep pace
tho' I rush to remain
in the very same place

One more long winter's over
the spring sun warms my back
while flowers and grandkids
keep life on its track

Let us shun major changes
and slow gently this way
then no one will note
when we drop out of play

For coming makes such a disturbance
that going should be a release
So head for the end very slowly
in order to reach it in peace

Death among the seagulls and salmon

I was getting the flu
and feeling quite old
when we took a slow walk
past a stream clear and cold

I had wanted to show her
where salmon touch earth
leaving oceans of plenty
for the stream of their birth

The seagulls were screaming
a black bear was in sight
all enjoying the feast
with no reason to fight

Long years had gone by
since I first saw this scene
when I came with my boss
to a bay quite pristine

The orders were given
my task was quite clear
"Count salmon that enter
each day through the weir"

Many trees quite enormous
on rocky coasts grew
where seals barked fish stories
some doubtless untrue

Those bear trails were handy
through forests of yore
when clam tides went crawling
far down on the shore

All the bears I encountered
stepped kindly aside
for with salmon so tasty
and more on each tide

To argue was pointless
just a waste of good time
As we went on our way
snorted greetings were fine

The boss died years later
for reasons unclear
leaving life quite unchanged
for both eagle and deer

Now it's my turn to wobble
while youngsters jog past
though I try to keep moving
each step seems my last

Quite plainly these fish
and the gulls that they meet
would feel right at home
with the first I did greet

As their ancestors too
fought to claim the best space
so descendents anew
might arise in their place

In life's play most intense
all the actors seem new
as I totter on back
from a stream I once knew

****

Enough beats "Too much"

A house that warms in winter
yet opens wide for spring
a summer house and loving spouse
is there some better thing?

A harvest house with food to share
where children like to be
Some space for books in quiet nooks
where conversation's free

Yet many who could have all that
prefer to seek more power
their lives on hold while assets cold
make money every hour

But all need love and love's a thing
that's neither bought nor sold
for love takes time and time runs out
on even wealth untold

Sweet time that's spent in discontent
can never be repaid
While others try to rule the sky
I'd rather just get laid

****

Pascal's "Why not believe?" formula couldn't cope with eternity

The argument summarized:

Humans have invented countless religions to worship innumerable gods. ***Religions come and go, and their theologies***

differ wildly. *No religion has ever had convincing evidence to support its magical claims.* ***So quite sensibly, most religions insist all the others are false.***

If we could grant every religion ***an infinitesimal chance*** *of being right - and each was then* ***tested for an infinite length of time - even the least likely religious proposition would eventually come true.*** *But eternity is not relevant to this universe or to ordinary human affairs, so implausible* ***religious claims are best treated as impossibilities.***

In contrast, the regularly demonstrated power of evolution theory arises from very large real numbers. Furthermore, unlike religious promises - which for practical purposes come true "only in your dreams" or "when hell freezes over" - the credibility of evolution theory does not depend on eternity or infinity. For the random processes that drive life's evolution operate through innumerable concurrent competitions between the consequences of countless unlikely events taking place in ordinary time.

A modern version of **Pascal's wager** goes "***Why not believe in God*** (or Heaven or Jesus or Mohammed or Christian Science or souls or the Devil or the saints or the prophets or Billy Graham or the Ayatollah Khomeini or prayer or human sacrifice or witchcraft *or whatever)?*

After all, what have you got to lose?" Obvious answers include: *Time. Money. Freedom. Independence. Privacy. Intelligence. Self respect.* ***In other words, your entire life!***

Yet Pascal was not just another religious con man selling snake oil. Rather, he was an accomplished mathematician and philosopher who ruminated obsessively about religion in general and Christianity in particular. Furthermore, Pascal's wager on the existence of God represented a legitimate mathematical effort to evaluate religious probabilities.

The fact that mathematicians of Pascal's time (1623-1662) were not equipped to handle infinities would have been irrelevant for determining the odds in any ordinary wager. But once Christianity included *eternal* happiness or *eternal* torment as part of its definition (though an extremely unlikely claim at best) - and once Pascal, for the sake of his mathematical argument, accepted that definition - the outcome was inevitable.

One might reasonably contend that Pascal would never have accepted Christianity had it merely guaranteed six hundred and sixty-six years of uninterrupted bliss, since the equal improbability of that claim was not overwhelmed by infinities. Regarding mathematical judgements, William James once wrote "they are all 'rational propositions' ... for they express the results of comparison and nothing more. The mathematical sciences deal with similarities and equalities exclusively, and not with coexistences and sequences."

Point 1) Incredible claims (religious or otherwise) only become credible when supported by overwhelming evidence.

Point 2) A mathematical calculation of potential yield has often been mistaken for overwhelming evidence.

****

May the odds be with you!

When Pascal, the math whiz,
saw himself growing old
he gambled on Heaven
o'er a grave still and cold

With his formula strong
and mathematics most rigorous
he figured the odds
on church tales ambiguous

Though likelihoods small
gave little confidence
when tied to forever
they made far more sense

So the obvious choice -
that death is the end -
was forced to give way
to infinity's trend

Since Pascal, the math whiz,
was undaunted by facts,
on that formula strong
he did base all his acts

And when death finally came
Pascal quickly did call
"Eternity's infinite
which outweighs them all!"

But if you become trapped
between fact and belief
and the latter doth promise
a whole lot less grief

you might disregard choices
that seem quite untrue
Let the life that you lead
be the best you can do

For why base your life
on a sale incomplete
when no guarantee holds
that the contract they'll meet?

Would you give all your cash
to a car salesman clever
for an auto unseen
that should last you *forever?*

Far better to purchase
the old Brooklyn Bridge
at least you can cross it
to reach the far ridge

Then return as you like
to seek those that you paid
if the product itself
is not quite as portrayed

****

Kindly skip the funeral

Lord save me from all the church ladies
that gather like flies when death's near
each hoping to test strange convictions
and prove there is nothing to fear

They claim we are simply God's children
and sigh "it was all for the best"
so surely I cannot be angry
"since living is only a test"

Lord save me from those who once knew me
now seeking some way to express
how their prayers, so far, have been answered
while I must go under duress

They rise to speak willy nilly
each sobbing that he loved me best
if only I'd lived the way he did
I might not have failed on this test

Lord save me from those second guessers
who distance themselves from my fate
denying what we have in common
until it's their turn and too late

And damn all the others most righteous
each fool who thinks You know what's best
for confounding young minds with such nonsense as
"Living is only a test"

****

God's three places

Old maps of the world
displayed many blank features
wherein mapmakers placed
the most fabulous creatures

Beyond here be dragons
was meant to imply
that those traveling farther
were likely to die

With the edge of the world
being rumored quite near
a fall o'er that edge
was the source of great fear

But explorers intrepid
sailed farther afield
and tales most horrendous
were slowly revealed

To be old myths recycled
by those staying home
and not the true stories
of men who did roam

Soon many survivors
reported their finding
just continents and seas

round the whole Earth a'winding

Now the world's many maps
are all filled in and boring
And no place remains
for those sea monsters roaring

Thus old myths must die
as science fills in the gaps
with satellite nav
replacing old maps

In the same simple fashion
were solved many mysteries
that once lent great power
to the holiest histories

With just three mysteries left
for science to banish
some fear that God's rule
o'er this life will soon vanish

"Hold on! Not so fast!"
the creationists cry
"I would rather lose science
than see my God die!"

"For by exposing the truth
that God's never been needed
science leaves us no Leader
who has to be heeded"

Do not worry
the scientists blithely declare
You still have three mysteries
and God could be there

So in what secret places
might God's touch remain?
The big bang. And life's start.
And your self-conscious brain

Well, the universe inflated
in far less than a day
arising from nothing
a *big bang* so they say

Though it might have begun
with an Almighty Kick
to keep it all running
appears no great trick

But if Great God in truth
gave this Universe the Boot
for ten billion years
perhaps craps He did shoot

Till the moment arrived
for Him to make life.
A brief intervention
that caused endless strife

Four billion years later
we encounter Him lastly
installing sweet spirits
in humans most ghastly

Perhaps consciousness human
was God's great reward
since a worshipful few
could then live by the sword

And call out His name
as their battlefield cry
while they rape, loot and shout
Non-believers must die!

Of course, nobody knows
just what consciousness is,
some take this as proof
the design must be His

Human souls might exist
though they've never been sighted
which seems to suggest that
they ought not be slighted

Perhaps each has a presence
Let's define this as spiritual
so science can't disprove it
with evidence empirical

But with three mysteries left
that Great God might reveal
a fourth mystery rises
Is God even real?

Since no one has met Him
nor gazed on His face
such tales from the Stone Age
now seem out of place

For why would Great God
just be hanging around
with His Universe on idle
not making a sound?

Simply watching what happens?
How eternally boring!
Except for those times
He could come out a'roaring

And scare the be'jeesus
out of all His Creations
Thus keeping them safe
from all thought and temptations

If this tale so extended
to you makes no sense
perhaps you'll start thinking
and stop the pretense

That God is in charge
of the life that you lead
that He will reward you
in spite of your greed

So if now and again
you would seek a solution
just sit in your chair
and review evolution*

(*Editor's suggestion - you might begin with Human Evolutionary Biology by *Arndt von Hippel* pub. 1995 by Stone Age Press, 548 pp. $29.95)

****

Heaven(ly?)

Religions teach what ought to be
and claim what might have been
To point out contradictions there

has always been a sin

A human mind thinks countless things
and often gets them wrong
tho' complex edifice be built
it rarely lasts for long

Yet some still stand as stories grow
with explanations varied
then others join for who's to know
wherein the truth be carried

When more agree a tale is true
that makes it quite impressive
let's all bow down to worship God
and hope he's not aggressive

With troubles spread this world around
no need to make Him mad
so sing His praises loud and clear
to show Him you are glad

Though inside all the doubt remains
Who'd make a world this way?
Till Heaven's gate we'll have to wait
at least that's what they say

And so, for now, let's pray anew
that troubles pass us by
and land on someone we don't like
to prove that God is nigh

****

Ethics and moral behavior develop through open competition and natural selection. Monotheistic fundamentalists routinely depend upon unethical and immoral behavior to gain and retain totalitarian control.

Every organization has rules of conduct to regulate relationships amongst its members. Harsh selective pressures often force closely related individuals to cooperate and even sacrifice for one another. While a sacrifice made by one person to benefit another might seem foolish, such altruism can actually pay long-term rewards if it leads to the production or survival of more (and more closely related) descendents.

We tend to admire cooperation and sacrifice. In contrast, antisocial behavior may pay short-term dividends for the bank robber or cancer cell, but it usually fails to bring lasting advantage or enhance reproductive success.

An interesting computer challenge has clarified some aspects of how cooperation (reciprocal altruism) can benefit unrelated individuals - even though natural selection might appear to favor purely selfish behavior during such encounters. Our simplified example includes a buyer, a seller and an exchange of packages. One package allegedly contains certain goods while the other package allegedly contains an agreed upon payment, with packages to be opened later.

The options allowed were cooperation or cheating. The winning strategy of each round was the one that maximized wealth through buying and selling. Many complex programs were submitted but as the number of transactions increased, only a simple program called *Tit for Tat* remained consistently among top winners. That program merely directed its player to start honestly and thereafter respond in kind (either honestly or dishonestly) according to the previous move of its opponent.

Because Tit for Tat did not initially seek to exploit, it suffered early setbaçks when competing with a large population of selfish participants. But a turnaround occurred as the population of suckers diminished, since exploiters were unable to do business with each other once their reputations became established. Reciprocators then flourished on the benefits of cooperation.

However, if the contest continued - and occasional errors allowed these competing programs to evolve - a more generous, even more successful version of Tit for Tat developed that forgave the occasional lapse in order to retain the benefits of cooperation. A recent entry may be even more effective because its flexible *Win Stay, Lose Shift* strategy can prevent all trading from coming to a halt - as when Tit for Tat stops trading because only cheaters remain. Under such circumstances, Win Stay may shift tactics in order to benefit from continued trading.

So even amongst computer programs, a reputation for rectitude, cooperation and reciprocity appears to be important for successful long-term relationships. Society, too, may benefit by forgiving the occasional transgression rather than always insisting upon getting even. For sometimes it is just not practical to be perfectly honest or completely fair, or to drive within the speed limit. And in order to survive, we have all learned to respond in Tit for Tat fashion (or its more generous version) as well as with Win Stay, Lose Shift behavior.

Cooperation arises naturally at home where most early dealings are with relatives. Except for that behavioral start, however, different groups tend to view morals and ethical conduct quite differently, depending upon whether they are headhunters or Quakers, lobbyists for a tobacco firm or Amish farmers, members of a large corporate hierarchy or Iranian Mullahs, radical environmentalists or anti-abortionists.

Evidently, the definitions of moral or ethical behavior vary so widely because - rather than being universal truths as so often portrayed within any group - they merely reflect the dominant or leadership consensus on goal-oriented rules that ought to govern social conduct within and outside the group. In most societies, ethical conduct includes any behavior that is regularly accorded social approval because it benefits more leadership interests or related individuals than other adequately tested behaviors.

The above considerations may partially explain why ***religious exclusionary behavior*** *(Treat co-religionists within* *the group as you would like to be treated, but feel free to convert, cheat or otherwise harm all outsiders)* ***and other forms of tribalism are such tough habits to break.***

Nonetheless, religious intolerance and other unethical and immoral tribal behaviors usually decline during good times when those primitive and unkind support systems seem less essential for survival - which suggests that you will personally be better off when most others become better off as well.

****

Nights alone

All the stars bright tonight seem to shrivel your might
though your worries just keep on a'growin'
still it doesn't take long for an ego not strong
to resume all its puffin' and blowin'

Many thoughts too intense for a night so immense
may seem scary while you are alone
but they won't be that way when the very next day
you're too busy to talk on the phone

The problem may be that the farther you see
the harder it is to be sure
And as things you once knew take a feeling untrue
it gets tough to tell steer from manure

So this lesson I've learned from those nights I have yearned
just give thanks for the stars in their glory
for the truth lies out there and since no one knows where
each can make up his own favorite story

****

Outcomes

Most churches teach their Stone Age truth

and try to teach it well
and if you don't believe, forsooth,
then you can go to Hell!

But those that sell the future
want their payment in advance
Since each describes it differently
you're taking quite a chance

Religions rest on many myths
all claim just theirs are right
and if you try to disagree
there'll surely be a fight

For churches cannot stand new facts
that progress would allow
since Stone Age science at its best
could never tell you how

to build a plane or make a car
or drill for oil, in truth
Yet myths they sell, and packaged well,
to clog the minds of youth

Christ's churches aim an awful curse
at all who seek more learning
For apple chewed by humans nude
in Hades you'll be burning

But primates testing heresies
by burning at the stake
are most concerned that aught be learned
so doctrine doesn't break

For churches just rewrite old news
(new facts they cannot fake),
then hope and pray each Sabbaday
their Stone Age myths still take

****

It was wrong to shoot the pope

Some Protestant leaders said it was no bad thing when the pope was shot. Some Catholic clergy wrote that it was no bad thing when the abortion doctor was shot. The Oklahoma City bomber had his defenders, as did recent murderers of gays and blacks. In each case, those of one minority were eager to excuse illegal violence against members of another minority - not a bright idea.

We all know how easy, exciting and initially rewarding it can be to generate hatred and irrational behavior - as in Northern Ireland, Bosnia, Germany, Rwanda or Israel. But ours is not a homogeneous society either. Indeed, every American belongs to some racial, religious or other minority. And the great strength, stability and productivity of our society clearly depends upon good will, opportunity for all and a tolerance for diversity.

Religious fundamentalist leaders regularly enhance their own position through fiery condemnations of legal beliefs and legal behaviors tolerated by other minorities. Those who recklessly incite the crazies of their own faction in this way are as responsible for any injuries that follow as those who falsely shout "Fire" in a crowded theatre.

****

Redemption

Beware of the true believer
who so willingly kills for his God
for when actions insane cause disruption and pain,
then religion seems other than odd

Beware of those offering redemption

as they take away all that you need
they may pray for your soul but their principal goal
is the satisfaction of greed

Beware of those seeking great power
who promise to help you the most
for by means somewhat stealthy they soon become wealthy
while their promises turn into toast

You may counter with true information
but facts cannot carry the day
for faith gains more power during every long hour
that those in the foxholes still pray

And each who comes home with the living
knows the Lord stood beside him all night
yet no one can tell why those other men fell
God-forsaken amidst their last fight

With the rules so unclear and the sales pitch just *fear*
what separates *saved* from the *damned?*
One is forced to conclude, though it may well be rude
that no fate has ever been planned

****

Every Sunday we pray

Every Sunday it happens
throughout our great land
many people do gather
and together they stand

Their children fresh scrubbed
and in clothing quite neat
each tries to pretend
that the church is a treat

While singing loud praises
so God will them favor
and thinking that breakfast
in bed they would savor

The children see sunshine
and wish to go play
The preacher is boring
with nothing to say

He speaks very kindly
in cliches and half truths
and treats each adult
as if one of his youths

Yet some he must frighten
before their donation
so he shouts of Hell's fire
and threatens damnation

Many harbor great doubt
'bout such mythical stories
then look all around
at those singing His glories

And marvel anew
at the power of prayer
to convince those in doubt
that their God might be there

The preacher at night
all alone in his room
sees profit a'plenty
in tales of great gloom

Used cars he once sold
now its on to insurance
as he tries to convince
through both guile and endurance

That when Heaven's at stake
'tis far better to pay
than a reject to be
on that great Judgement Day!

Though it's hard to believe
in a bliss quite eternal
One cannot take risks
with Hell's fires infernal

Each thinks that the others
don't doubt those Flood stories
Yet everyone shares
the same secret worries

For how could a whale
know that Jonah lacked fault?
and why turn Lot's wife
to a pillar of salt?

Thus it doesn't make sense
that all these fine people
should gather together
beneath a large steeple

To worship strange beings
made up from the air
and speak of lost souls
roaming no one knows where

Yet if souls all be doomed
till they pay to reach glory
and no souls have been seen
might this be just a story?

****

Sex and the spider

The spider weaves a complex web
across my door each day
She hopes to catch me for a meal
When I go out to play

The spider's eyes see many things
but must not see them clearly
for her tummy small, with prey so large,
would overstretch severely

The spider cannot stretch her legs
with strong extensor muscle
instead she uses fluid drive
to help her in a tussle

The spider is an ugly beast
whose bite dissolves your meat
She always drinks her insects straight
and views them as a treat

The spider eats her mate for lunch
He gets a rush, they say
I think I'd rather skip the sex
and live another day

****

Just another Stone Age fable

Here is what happened
there's no room for doubt
all was in chaos
'til God gave a shout

Our Father decided
He knew what to do
Make daytime and night-time
Add sunshine. Then brew

Make trees, fruits and flowers
pussy willows galore
make frogs, snakes and fishes
who knows what's in store

Make animals countless
and all good to eat.
Make man in His image
and woman so sweet.

Let Adam be stupid
unable to think.
Then Eve can delight him
with fruit and with drink

All seemingly perfect
though puzzling as Oz
for brains were unwanted
under Eden's strict laws

Then along came a snake
who spake ever so clearly
A bite of this apple

may cost you quite dearly

But the flavor's delightful
and it will explain
why God in His heaven
made you such a brain

Well an order's an order
when you're perfectly stupid
so they ate of that fruit
and discovered old Cupid

Soon Adam eyed Eve
with a look so lascivious
that she covered herself
with a fig leaf herbivorous

Then Adam complained
Your fig leaf's distracting!
In fact, he declared
I'm thinking of acting!

Eve became worried
Can't you wait till tonite?
For I'm terribly scared
this will cause a big fight

But poor Adam could see
just six inches ahead
till God came upon them
Then both feared they were dead

So Adam confessed
it was all Eve's mistake
And the snake too had caused them
their promise to break

But now they both knew
life had far more in store
than smelling the flowers
or opening God's door

Soon with a thunderclap
terribly loud
they were cast from God's garden
in the midst of a cloud

Oh, the evil let loose
by that Original Sin!
From fat, dumb and happy
to outside looking in.

Though losing pet status
in Eden seemed wild.
Eve was soon feeding
their number one child

Natural childbirth so painful
by God's own decree
and weeding back-breaking,
only Eden stayed weed-free

Yet eating that apple
left more punishment due
which might seem quite silly
to me and to you

But God was a stickler
on how Eden should run
To prove this was true
He soon killed His own Son

First God became bored
with the Earth He'd created
and decided to drown
all the life He now hated

When Noah (God's buddy)
heard flood times were due
he invested in gopherwood
for an Ark big and new

Old Noah was wise
in his choice of a crew
just wife and descendents
no wages were due

God's creatures then boarded
each two and by two
along with some extras
to put in the stew.

Noah chose elephants,
tapeworms and rats
God spake unto Noah
Don't forget to bring cats!

Of dinosaurs, God said
I will miss them the least
since ne'er did I make
a less charming old beast

They are noisy and smelly
and if they do play
the ground shakes and trees break
when they get in the way

Well the rain fell in torrents
(from God's buckets they say)
and the sea rose right swiftly
100 meters per day!

But Noah's fine gopherwood
Ark so immense
leaked nary a drop
if that makes any sense

While the animals all
in a manner so kind
each cleaned its own stall
with no need to remind

For they knew the Ark built
for passengers lucky
had been very hard work
for Noah's family plucky

So that's how it happened,
creationists insist
and if you show doubt
then they really get pissed

Since their story reports what
some Stone Agers' knew
A tale once convincing,
tho' long deemed untrue

Still this myth may sustain
when you feel like you're sinking
some lose very little
by giving up thinking.

And for them Eden stays
such a fine happy place
where all of God's children
gather flowers and embrace

All innocent and pure
with no one to defile
since nowhere in Eden
was sin e'er in style

But flowers are
sexual structures too
like the birds and the bees
they exist just to screw

Wherein lies a great lesson
to which you should heed
birds sing and bees work
and all screw to succeed

And while you may cherish
tales old and untrue
only science kills germs
and makes microwaves too

Perhaps faith does move mountains
when no one is looking
but electricity's handy
when time comes for cooking

So believe as you like
but let others learn well
for the Stone Ager's life
was a version of Hell

And if for your Faith
you attempt to kill knowledge
you may live to regret what
your child missed in college

Since wanting to know
was the very first Sin
thank God for the apple
that let thinking begin

And here's to the snake
who brought all this about
who'd have dreamt a mere serpent
would have so much clout

As for me, I am happier
thinkin' and weedin'
than I ever could be
in the Garden of Eden!

****

The Holy Trinity uncovered

The ancient Celts were excellent fighters, according to both Greek and Roman reports. Although Celtic battle tactics remained simple, their soldiers had a dominating presence and were renowned for fierce attacks that they carried out amidst a deafening din of trumpets and battle cries. Interestingly, Celtic warriors always fought naked, presumably to enhance their power and success in battle. Other possible explanations - Celtic warriors were exhibitionists - they were extremely warm-blooded - they rarely caught colds - have little explanatory value, especially in winter.

Among notable religious events of the fourth century AD were two conclaves by church authorities. At the first, in Nicaea during the year 325, a simple majority decided Jesus Christ had been God all along - not just another holy man, teacher or prophet. That the church delayed three long centuries after His death to make this determination seems a bit much, even for a widespread bureaucracy, since by then the few known facts about Jesus had vanished into a sea of speculation and myth. Nonetheless, monotheistic Christians could worship their newly doubled God without major upset. For double gods were still quite common in those days, and Celtic pagans sacrificed to single and three-in-one gods as well.

To view Nicaea in its proper historical perspective, imagine trying to get a majority of the current U.S. Congress, two centuries after George Washington's death, to agree that our first President really had two heads. Of course, unlike the presumably perfect life led by Jesus - about which so little is known - President Washington's long life and career were carefully recorded by many of his contemporaries. In any case, electing Christ to the Godhead seems to have caused sufficient doctrinal difficulty that a follow-up meeting had to be called at Constantinople in the year 381. And it was there that church authorities elevated a mysterious Holy Spirit or Holy Ghost to the Godhead as well.

By enlarging the recently doubled Christian God another 50% to create the One True Three-in-One, Constantinople made Triple-Him seem even more radically different from the earlier one-and-only Jewish God or the later one-and-only Muslim God. Yet from whence this Holy Ghost arose, and how It/He suddenly came to be recognized by majority vote, was never made clear. Nor was this decision left open for further debate by either Catholics or Protestants, as Michael Servetus learned when burned at the stake in Geneva (1553).

In 1676, Anthony Sparrow, Bishop of Norwich, placed himself heartily in favor of celebrating a special annual Trinity Sunday, declaring "that such a mystery as this, though part of the meditation of each day, should (also) be the chief subject of one. ...for no sooner had our Lord ascended into heaven, and

God's Holy Spirit descended upon the Church, but there ensued the notice of the glorious and incomprehensible Trinity, which before that time was not so clearly known." In other words, he didn't get it either.

Nowadays, the Holy Ghost is usually passed over lightly as an especially mysterious aspect of God's ultimately unknowable essence. However, there may have been a far simpler, more down-to-earth reason for concocting a Holy Spirit during that Constantinople gathering. For It/He was added to the Godhead at a time when intense missionary work was being directed toward the Celts. However, those Celts already had their own triple deities, and Celtic religious leaders insisted that a triple god had the most clout. Consequently, the One True Christian Double-God would have been a hard sell to Celts, even if backed by a victorious Roman army.

Thus the Holy Ghost may simply have been added to the Christian pantheon so that the resulting tripartite Christian God would resonate with the large Celtic populations of Europe. But in order to support or reject such a radical conclusion, we must first determine why Celts felt triple gods were so all-powerful. Well, we have already referred to the ancient Celtic belief that exposing the basically tripartite unity of their male genitalia in battle somehow enhanced their fighting power. And surely, a merely bifid Christian God would have brought to mind the far less powerful female genitalia, thereby making Christianity an object of Celtic derision rather than devotion.

Under these politically urgent circumstances, it made little difference that the Father/Son/Holy Ghost was a relatively late (nearly mid-millenial) and purely Christian creation. Nor did it matter that the Trinity was never once mentioned in the entire Hebrew Bible, or in the New Testament. And few even seemed to care that Jesus Christ Himself appeared unaware of this interesting conglomeration. When seen in that light, the Holy Ghost simply reveals an implicit but never before mentioned aspect of the all-powerful - hence surely tripartite - Christian God. So that Holy Spirit is best explained as a stand-in for His left testicle.

Does any other evidence favor our hypothesis about the mysterious Holy Spirit? Well, until recently all Christian worshippers were divided into two classes, based entirely upon the appearance of their external genitalia. Those possessing powerful tripartite genitalia formed the upper or leadership class, while anyone suspected of having bifid genitalia was automatically relegated to the non-voting lower class or service category.

In order to drive that point home more emphatically, some early Christian church steeples had a powerful resemblance to an erect phallus. And to this day, true steeples rarely if ever arise from the center of a church. Indeed, steeples only look "natural" when located near one end of a wider, often somewhat asymmetrical structure within which parishioners are temporarily stored until emotionally pumped to spread the word.

This brief discussion has obliquely considered the impact of religious politics and strategy - and the churchly desire to retain authority and control - on the formulation of Christian dogma. Future historical research may reveal that successful churches have rarely if ever made major doctrinal changes solely in response to heartfelt religious concerns.

****

The Holy Trinity (part two)

In olden days when Celts attacked
and raised a fearful din,
with trumpets blare and battle cries
quite often they would win

no clothes they wore when arms they bore
for warfare was no game
their manly power made strong foes cower
and earned the Celts great fame

These warriors bold were never cold
their children too stayed brave
though battles won in rain or sun
sent fathers to their grave

The Celtic gods liked warfare best
and human parts to eat
the strongest gods had triple heads
all three demanding meat

The Christian God so infinite
His head was one of note
till Christ's own head was added
at Nicaea by a vote

T'was there in year three twenty-five
the churchmen met and ruled
that God with Christ wore two hats too
no longer were they fooled

Though centuries past had left no trace
of Christ's own life so pure
the Nicaean vote set Christians straight
and banished choice unsure

Now Christ with God meant double heads
the whole world would then follow
even Celtic men might join a faith
that beat their own so hollow

But Celts just laughed at Christian claims
of doubled-God's great clout
for Celt tripartite gods were males
who bifid gods would rout

Thus Christians had to meet again
in year three eighty-one
the Holy Ghost they voted in
to make God three-for-one

The Celts all knew what triples do
with those they take to dinner
it made the Christians very glad
that now they had a winner

The Holy Ghost was hard to see
with meaning undeclared
for those who failed to get the point
large fires were prepared

Some Christians pure remained unsure
would God's crown fit the Trinity
since Bible never claimed a Ghost
nor Christ one-third Divinity

Soon Celts did join the Christian faith
St. Patrick spread the word
for power of three as all could see
each soldier's loins did gird

In Christian churches to this day
tripartites hold all power
while bifids of the lower class
serve coffee every hour

It might seem crude but I conclude
old steeples round were phallic
So if doctrines change in ways most strange
just write *MARKETING* in italic.

****

Unwed fathers

Unwed fathers, all is forgiven
the fault was not yours in the least
for those promises made and the fair maid betrayed
She should have known man was a beast

Nobody saw you. She's clearly immoral
Her word against yours has no prayer
So who cares what they think as with others you drink
while she tries to arrange for child care

Unwed fathers, everything's changing
Anonymity lost on the day
that you spawned a new child on an evening so mild
for that son bears your own DNA

With science progressing, it is truly a blessing
to establish each parent for sure
Now the radical right may soon try with their might
to promote birth control or stay pure

New times are confusing, it's unclear who's using
the other when games change this much
So be friends and take care that you don't leave an heir
when you'd rather remain out of touch

****

Virtual reality

If wishes were Hondas
all the beggars would ride
Every vegetable sold
would have caramel inside

Gas would be free
and pollution unknown
cholesterols normal
your children all grown

Grandchildren happy
would rush to your call
and each would declare it
no trouble at all

Cows would lay steaks
just as hens lay their eggs
And people would run
on untiring legs

Of time there'd be plenty
with extra to share
when you and your sweetie
went off to the fair

Learning would be easy
and people all young
The crippled could dance
till the last song was sung

The blind would see brightly
all night and all day
the mute would sing sweetly
on harps they would play

Now the moment arrives
to insert your next dollar
or go out in the rain
Will you raise up your collar

and limp along home
to a room by the tracks
near a john that won't flush
and a wall full of cracks?

Or spend your last buck
on this world you've confected
where all troubles are banished
and life's been perfected?

****

About the Author

Arndt von Hippel was born in Germany in 1932 and arrived in the United States in 1936. He comes from a unique multigenerational scientific family that was featured in *The Scientist.* Dr. von Hippel acquired his B.S. in Biology from M.I.T. and an M.D. from Harvard. In 1965, after eight years of surgical training, he moved to Anchorage with his pediatrician wife and their children, where he opened a solo practice in chest surgery and later developed Alaska's very successful heart surgery program.

Following retirement, von Hippel taught a popular course in human anatomy and physiology at the University of Alaska. Because none of the available textbooks explained human anatomy and physiology from an evolutionary perspective, von Hippel wrote his own widely respected *Human Evolutionary Biology* book (548 pp, pub by Stone Age Press, Anchorage, 1995, $29.95). And now, in response to all those poorly informed, premillenial creationist attacks on Evolution Theory, von Hippel has provided us with *An Evolutionist Deconstructs Creationism.*